MOLECULAR STRUCTURE

Macromolecules in Three Dimensions

MOLECULAR STRUCTURE

Macromolecules in Three Dimensions

Robert J. Fletterick
Department of Biochemistry and Biophysics
University of California
San Francisco

Trina Schroer
Department of Biochemistry and Biophysics
University of California
San Francisco

Raymond J. Matela
Faculty of Technology
The Open University
Milton Keynes, Great Britain

Blackwell Scientific Publications
Oxford   London   Edinburgh   Melbourne
Palo Alto   Boston

Editorial Offices

706 Cowper Street, Palo Alto
California 94301, USA

Osney Mead, Oxford, OX2 OEL

8 John Street, London, WC1N 2ES

23 Ainslie Place, Edinburgh, EH3 6AJ

Distributors

USA and Canada
Blackwell Scientific Publications
P.O. Box 50009
Palo Alto, California 94303

United Kingdom and Europe
Blackwell Scientific Publications, Ltd.
Osney Mead, Oxford, OX2 OEL

Australia
Blackwell Scientific Book Distributors
31 Advantage Road, Highett
Victoria 3190

Editor: John Staples

Production Coordinator: Deborah Gale

Artists: George Klatt and John Waller

Interior and cover design: Gary Head

Composition: Jonathan Peck Typographers, Ltd.

Printing and Binding: Maple-Vail Book Manufacturing Group

Mold Maker: Bruce G. Argetsinger

First published 1985.

Blackwell Molecular Models, Patent No. 4378218

Library of Congress Cataloging in Publication Data

Fletterick, Robert J., 1943–   Molecular structure.

   Includes index.
   1. Macromolecules.   2. Molecules—Models.   3. Proteins.
4. Nucleic acids.   I. Schroer, Trina, 1959–
II. Matela, Raymond J., 1946–   .   III. Title.
[DNLM: 1. Macromolecular Systems.   2. Models, Molecular.
3. Nucleic Acids.   4. Proteins.   QU 55 F615m]
QP801.P64F55   1985   574.8'8   84–12501
ISBN 0–86542–300–8

Preface

Exploring the three dimensionality of our world has been the subject of Max Escher's artistic work. He was, as we are, tormented and fascinated by the complexity of three-dimensional images. He explored the confusions and illusions with compelling works such as the one on the next page.

Escher wrote:

Our three-dimensional space is the only true reality we know. The two-dimensional is every bit as fictional as the four-dimensional, for nothing is flat, not even the most finely polished mirror. And yet we stick to the convention that a wall or a piece of paper is flat, and curiously enough, we still go on, as we have done since time immemorial, producing illusions of space on just such plane surfaces as these. Surely it is a bit absurd to draw a few lines and then claim: "this is a house."

Molecular structure is the basis of study in chemistry and biochemistry yet few students (and here we include professors of these subjects) can understand, recognize, or think about problems in three dimensions. The learning in these subjects is usually by two-dimensional representations, sometimes elegant but almost always inadequate because three dimensions cannot be represented in two without severe loss of spatial character.

Reptiles by M. C. Escher reproduced courtesy of © BEELDRECHT, Amsterdam/V. A. G. A., New York, Collection Haags Gemeente-musem—The Hague. 1981

Not long ago we invited some of our friends and colleagues to identify a backbone model of myoglobin that was newly constructed in Blackwell model parts. Myoglobin is a protein with a very characteristic structure.

Most knew it represented a protein model, and many recognized α-helices, but few knew it was a model of myoglobin and those who did thought that it was not the correct shape. Two of the viewers had published papers on structural studies of globins! Even professional structural biologists have difficulty identifying structure when confronted with a three-dimensional model. Ironically these same colleagues have little difficulty identifying structures of proteins from two-dimensional representations in backbone drawings.

The importance of understanding the structure of macromolecules in biochemistry has been underscored by textbooks that emphasize structural explanations of biological function. Although these books have been successful in their presentations of the modern concepts of biochemistry and molecular biology, many concepts that rely on three-dimensional understanding would be enhanced by physical models of some of the molecules. That is the gap which *Molecular Structure: Macromolecules in Three Dimensions* and the *Blackwell Molecular Models* attempt to fill.

Our discussion of the structure-function relationships in macromolecular structure is presented in the traditional manner of introducing the components—the nucleotides and amino acids—and exploring the structures and properties of polymers of each. The basic physical principles are given for folding these molecules into functional conformations. Secondary structures and the details of tertiary structures are discussed with a completed model and many associated figures as additional aids. The biological functions of macromolecules are related to their structures at the level at which we understand them.

We have included several figures in *Molecular Structure* without fully explaining their significance or developing the pertinent background material. These figures are intended to pique the curiosity of readers about certain topics and to enhance their thinking about the topics that are presented in detail. Students who are particularly interested should refer to the voluminous original literature on these subjects.

Macromolecules are discussed at a basic level in this book. The reason for this is that two-dimensional displays with adequate discussions already exist for most of the molecules discussed in this book. Even though we hope that the book alone adequately explains macromolecular structure without the construction or examination of the models, this would defeat our purpose. The models are simple to build and should add some fun to the learning process. They will add a new dimension of understanding about complex macromolecular structure and should make basic concepts in this field readily understood by most college students in science programs and even by advanced high school students.

Many people have contributed to this effort. Lubert Stryer helped considerably with the development of the text. Informative comments on the project were provided by David and Jane Richardson, Steve Sprang, Elizabeth Goldsmith, Sung Hou Kim, and Diane Colby. Michael Connolly developed the computer programs for the surface representations of DNA, insulin, and lysozyme. Bruce Argetsinger engineered the models. Ric Bornstein typed the many drafts of the manuscripts. Mike Levers patiently photographed and rephotographed the models. Sally Boyle, Phil Evans, Arthur Lesk, and Wayne Anderson contributed to several of the computer and line drawings. Adam and Martha Kudlacik helped in more ways than can be listed here and deserve special thanks for helping the world know of the project. Finally, our editor John Staples is credited for establishing and developing a project that seemed radical at the outset and enlisting Deborah Gale, Gary Head, and Georg Klatt—an incomparable production team.

Robert J. Fletterick
Trina Schroer
Raymond J. Matela

How to Use This Book

Most molecular biology books discuss the principles of molecular structure by presenting structure using two-dimensional figures. *Molecular Structure: Macromolecules in Three Dimensions* offers the reader the opportunity to construct three-dimensional models of both simple and complex molecular structures using backbone models (the *Blackwell Molecular Models*). However, this text is not designed to be used exclusively in conjunction with backbone model kits. The text is generously illustrated with two-dimensional figures that will help the reader visualize and understand the physical, chemical, and biological principles that are discussed. The model kit and text can be purchased separately or together. Ideally, each student should have their own model kit. Alternatively, the instructor may wish to purchase several kits for students to share in a classroom model-building session. Since some students may not wish to build molecular models, extremely detailed figures and photographs have been included to convey the three-dimensionality of molecular structures. Of course, a student's understanding of molecular structure is enhanced if they are able to examine a molecular model. This understanding is increased if the student is able to construct the model, but much insight into three-dimensional structure is gained simply by viewing a completed model. In courses in which students are not asked to build models the instructor may wish to construct a set of models for the students so that they may examine the three-dimensional structures for themselves. All students in molecular biology, whether beginning or advanced, should experience the wonder of molecular architecture by contemplating backbone models of proteins and nucleic acids!

Only enough model parts are provided to construct some of the structures described in the text. A list of the number of parts and the time required to build each model is in the Appendix. We also suggest certain combinations of permanently glued models that can be built with the model parts provided in one kit. Students should work in groups of two or three to construct all the models discussed in the text.

Although this book focuses on the construction of backbone models, it should be used in conjunction with other types of models. The concepts of the intermolecular forces that determine secondary structure are best visualized using all-atom models. Space-filling all-atom models allow understanding of the close-packed nature of the interior of proteins and nucleic acids. The instructor may wish to offer students an opportunity to construct or view all-atom models of simple secondary structure elements to illustrate these points.

A slide set accompanies this book for instructors who wish to develop certain concepts further. The slide set is ideal for use in a course in molecular structure or in the molecular structure section of a biochemistry or molecular biology course. Its emphasis is on illustrating the concepts of symmetry and secondary and tertiary structure in molecular architecture. The set features figures from the text, particularly the computer graphics images and drawings of the models. A manual describing each slide in detail is provided.

Contents

PART III: MODELS OF PROTEINS 119

MOLECULAR STRUCTURE
Macromolecules in Three Dimensions

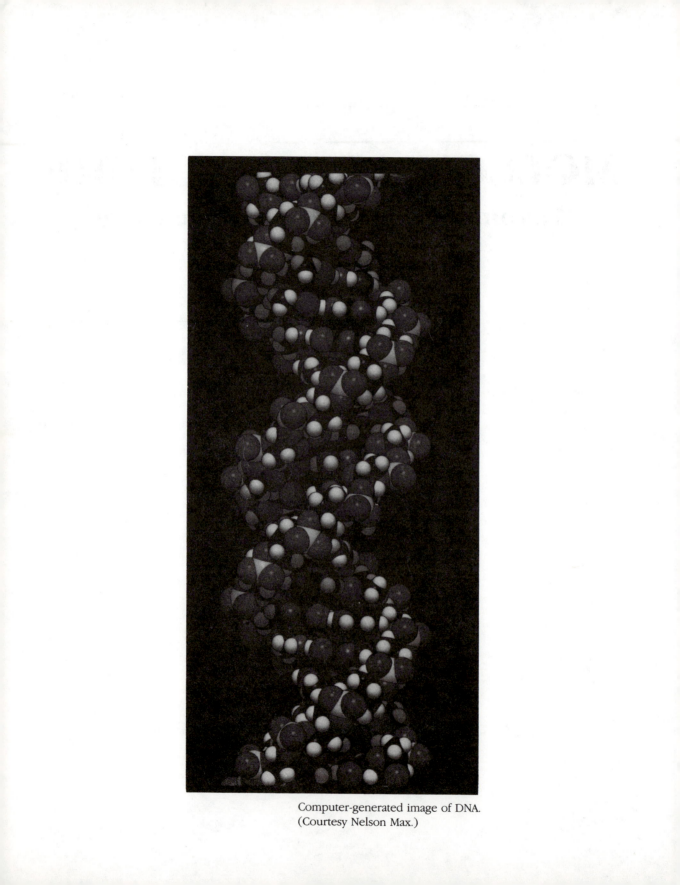

Computer-generated image of DNA.
(Courtesy Nelson Max.)

PART I

MOLECULES AND MODELLING

As "beauty is truth and truth beauty" so function is structure and structure function. Three billion years of biological evolution has resulted in the production of remarkable molecular structures that maintain and continue life processes. In Part I, we introduce you to *Blackwell Molecular Models* and other systems for examining molecular architecture. Common structural features of proteins and nucleic acids are also discussed.

1

Understanding Macromolecular Structure Using Models

The relationship between structure and function is so common and so obvious in everyday life that we seldom give it much attention. Usually familiar objects are, by virtue of their structures, specifically suited to their functions. It is especially true in biology (Figure 1-1). The relation between form and function extends to molecular biology, where a molecule's function is determined by its structure (Figure 1-2). In a very real sense, the complex and integrated reactions that occur in the cell follow from the structure of the individual components.

There are numerous examples of the contribution of structure-function relationships to modern molecular biology. One need only be reminded of the revolution that followed the elucidation of the structure of DNA (deoxyribonucleic acid); the discovery of complementarity in the double helix gave rise to more fruitful hypotheses and productive research than any other scientific observation in the 20th century. Similarly, solution of the structure of viruses and enzymes at atomic resolution has provided a deeper and more fundamental understanding of the way in which these entities function.

The three-dimensionality of biological molecules is difficult to envision, since they are usually presented, in textbooks, in two dimensions. Clearly, however, to completely understand the interrelation of structure and function, one must think of molecules in three dimen-

Figure 1-1
Darwin's finches are a classic example of the relationship of form and function. The different species have differently shaped beaks that suit their specialized feeding functions. The ground finch, *Geospiza magnirostris* (bottom), thought to be closer to the ancestral form, is a seed eater. The warbler finch, *Certhidea* (top), feeds on small insects in bushes.

sions. The most obvious way to present three-dimensional structures to students is to provide them with an opportunity to construct molecular models. Just as the appreciation gained by climbing a mountain is far different from that obtained by reading about it, actually building three-dimensional models provides a greater understanding of the structure-function relationship seen in complex biological macromolecules than could be obtained by other methods. Knowledge gained by experience is far more lasting than knowledge gained by acquaintance.

The purpose of this text is to provide a means for learning about the structure of proteins and nucleic acids by constructing simple models of each. We hope that an understanding of spatial relationships in molecules will follow from having the reader build models of them. We also hope that construction of the models and the accompanying discussion and thought will enable the reader to understand some functional aspects that follow

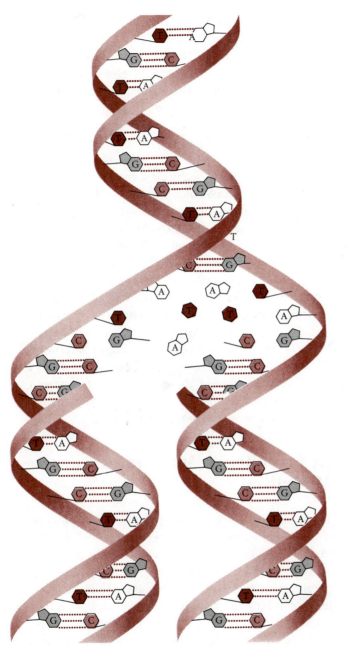

Figure 1-2
Replicating DNA. The two strands of the parental double helix unwind, and each specifies a new daughter strand by base-pairing rules.

from structure and then relate these concepts to other systems.

The specific projects described in this book are the construction of models of structural elements of proteins such as α helices and β structures as well as the construction of models of collagen, insulin, *cro* repressor, and lysozyme complexed with its substrate. Models of tRNA (transfer-ribonucleic acid) and the A, B, and Z conformations of DNA can also be built to show some of the varied conformations that nucleic acids may have.

Because these molecules are so complex, a system is required that allows the simple models to be constructed without the loss of too much three-dimensional information. The system that seems best suited for this is one that traces the backbones of these molecules through three-dimensional space, is rigid, and may be constructed in different colors. A further requirement is that the models be relatively simple to manipulate physically. These are the reasons why the *backbone modelling system* was chosen for both the nucleic acids and proteins. To clarify some important points that follow from the three-dimensionality of these molecules, the book has numerous two-dimensional figures that complement the completed three-dimensional models.

Approaches to Macromolecular Modelling

The backbone modelling system described in this book is not intended to replace other methods of modelling macromolecular structure. All-atom models are essential for research studies or for displaying details of hydrogen bonding and interatomic relationships. They are less useful for teaching the architectural principles of protein or nucleic acid structure and are too complex to be useful for comparing general features of protein or nucleic acid structure, even by professionals. The all-atom representation of one of the smallest proteins such as insulin has about 750 nonhydrogen atoms and 3×10^5 interatomic distances. The α-carbon representation is about 50 atoms and 1×10^3 interatomic distances. Visual complexity is reduced by a factor of perhaps $3 \times 10^5/1 \times 10^3$ or 300 in this case. Backbone models of proteins and nucleic acids are particularly useful for several purposes. Comparing complex structures and noting significant differences and similarities are much easier when one focuses on the molecular backbone. The use of protein backbone models allows study of the arrangement of secondary structure elements, and seeing the positioning of hydrophobic and hydrophilic residues is aided by the two-color scheme.

To examine the detailed atomic structure and particularly the hydrogen bonds that are critical to the stability of secondary structure in proteins and nucleic acids, one must use an all-atom modelling system. You may want to buy model parts to construct ten amino acids and four nucleotides so you can examine the internal packing of a segment of α helix, β structure, and a two–base pair stack in DNA.

All-atom models are useful for constructing short segments of secondary structures such as the α helix or a small segment of a DNA double helix. Large structures are expensive to construct and may even come to ruin with use and time. This is why most current structural research uses computer graphics to model protein and nucleic acid structures. Figure 1-3 shows a computer-generated model of the enzyme glycogen phosphorylase.

Computer graphics is undoubtedly the best way to look in detail at complex structure but is not yet inexpensive enough to be readily available for general use. This technology also suffers from the frustration of being "down the hall," "in another building," "in use," or "non-functional" just when it is needed to discuss a scientific issue. An additional psychological frustration is the inability to get a "hands-on" feeling with this technology. Most professionals who use computer graphics to study complex molecular structures also employ simple models in their research.

The only shortcoming of the backbone modelling system is the loss of detail. Fortunately, most of the detail is not necessarily of general interest, and the specific and interesting details may be discussed using conventional two-dimensional figures. These in turn have greater impact when inspected with accompanying backbone models.

Figure 1-3
Computer-drawn van der Waals surface of the enzyme glycogen phosphorylase. This is a dimer of identical subunits. The concave surface contains the active site and binds to a 300 Å particle of glycogen in the cytoplasm.

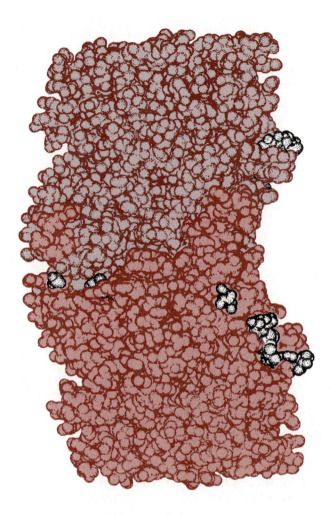

Using the Book and Models

This text can be used alone to learn about macromolecular structure, but the text's usefulness is enhanced when accompanied by a model kit. Students in many different courses, including protein or nucleic acid structure sections of an undergraduate or elementary graduate biochemistry course, will find the text and model kit useful supplements to a general biochemistry textbook. The text and model kit may also be used in the molecular structure section of a laboratory course. Backbone models are also an ideal accompaniment to a minicourse in protein or nucleic acid structure. Model kits are available from a number of sources, including Blackwell Scientific Publications, Inc., which distributes the *Blackwell Molecular Models.**

Since the models can be constructed rapidly, they can easily be assembled in a laboratory or class discussion section. The model of *cro* repressor protein can be built in three hours!

The *Blackwell Molecular Models* can also be used by students or professionals interested in building models of proteins other than those described here. This kit provides a review of the principles of macromolecular structure and an introduction to the main chain modelling system, which may be used to build any protein for which atomic coordinates have been determined.

Backbone models are widely used by professionals for research in macromolecular structure and function. Many university and corporate research laboratories have constructed protein backbone models, and a model of myoglobin is displayed in the Chicago Museum of Science and Industry.

The molecular models described in this book should be constructed using parts of two different colors (red and green). Models of proteins are displayed best with the hydrophobic amino acids a different color from the hydrophilic ones. The nucleic acids described in the text are color-coded according to symmetry, or in the case of tRNA, the two colors mark the special functional portions of the molecule.

We have described how to build many different protein and nucleic acid structures in this text. A photograph of each structure is included with the building instructions. You may not wish to build all the structures described here. In the Blackwell kit, 136 model parts are provided. These may be used to construct permanently glued models of one of the large protein structures or a few of the smaller structures. Additional parts and angle settings for nearly 100 proteins are available from Blackwell Scientific Publications, Inc., Box 50009, Palo Alto, California, 94303.

---

*Model kits can be ordered from Blackwell Scientific Publications, Inc., Box 50009, Palo Alto, California 94303.

Summary

Structure-function relationships are found throughout the biological world. Such relationships are particularly obvious at the level of atomic structure. Three-dimensional atomic structure is difficult to understand unless one can examine a three-dimensional model. We provide a unique three-dimensional modelling system and text that describes many aspects of protein and nucleic acid structure. The reader is given the opportunity to construct and study molecular models to learn about the basis and function of structure in biology.

Suggested Readings

General References

Cantor CR, Schimmel P: *Biophysical Chemistry,* Vol. 1. San Francisco, W.H. Freeman & Co., 1980. Rigorous discussion of proteins and nucleic acid structure and physical principles.

Richardson JS: Anatomy and taxonomy of protein structure. *Adv. Protein Chem. Rev.* 1981; 167–339. Excellent review of the architectural aspects of protein structure.

Schulz G, Schirmer RH: *Principles of Protein Structure.* New York, Springer-Verlag, 1979. Excellent source of references and discussion of physical principles.

Stryer L: *Biochemistry.* San Francisco, W.H. Freeman & Co., 1982. Excellent general discussion of basic principles of biochemistry.

Backbone Models

Fletterick RJ, Matela R: Color-coded α-carbon models of proteins. *Biopolymers* 1982; 21: 1000–1003. Description of the theory and features for the modelling system used in this textbook.

Models

Schultz GE, Schirmer RH: *Principles of Protein Structure,* New York, Springer-Verlag Press, 1979, pp. 134–144. Discussion of the many kinds of representations of complete structures of macromolecules. Examples are presented for each kind of display.

Atomic Coordinates

Bernstein FC, Koetezle TF, Williams GJB, et al: *J Mol Biol* 1970; 112: 535–542. Description of the Protein Data Bank, which is a computer data base containing atomic coordinates for all the published x-ray structural analyses of proteins and nucleic acids. Atomic coordinates for these structures may be obtained by subscription, sent to the address referred to in this publication.

Computer Graphics

Connolly ML: Solvent Accessible Surfaces of Proteins and Nucleic Acids, *Science* 1983; 221: 709–713. Description of the methods used for calculation and visualizing macromolecules. Color photographs are included.

Langridge R, Ferrin TE, Kuntz ID, Connolly ML: Real-time color graphics and studies of molecular interactions. *Science* 1981; 211: 661–666. Description of the use of computer graphics in modelling and understanding interactions of proteins and ligands.

Weiner PK, Langridge R, Blaney JM, et al: Electrostatic potential molecular surfaces. *Proc Natl Acad Sci USA* 1982; 79: 3754–3758. Description of the use of computer graphics to display the electrostatic surfaces of proteins and nucleic acids. Color photographs show how computer graphics can be used to understand charges on molecular surfaces.

2

Principles of Molecular Design

Figure 2-1
This crystal of glycogen phosphorylase *a* contains active enzyme. The crystal is approximately 2 mm long and is composed of 50% protein and 50% water.

To discover the salient features of macromolecular structure, one must examine molecules using optical methods. *X-ray diffraction* and *electron microscopy* are the primary techniques for determining molecular structure. Proteins and nucleic acids share many structural features. Both types of molecules are long polymers of repeating subunits. The way the long chains fold into secondary and tertiary structure is what determines the specific functional characteristics of each molecule.

Imaging Molecules

The structure of macromolecules must be examined in fine detail to determine the common principles that govern molecular architecture. Most of the three-dimensional structural data obtained for biological macromolecules comes from x-ray diffraction experiments or the electron microscope. X-ray diffraction analyses require that molecules be ordered in either fibers (one-dimensional), sheets (two-dimensional), or crystalline arrays (three-dimensional).

The x-ray diffraction analysis of a single crystal of a protein (Figure 2-1) begins with accurate measurement of the diffraction pattern (Figure 2-2). From such patterns it is possible to compute an electron density map and fit

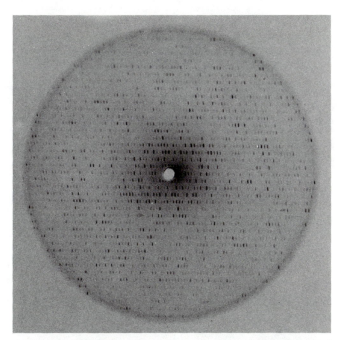

Figure 2-2
Single plane of the three-dimensional x-ray diffraction pattern from a crystal of the enzyme aspartate aminotransferase. In this case, a complete three-dimensional diffraction pattern would be composed of about 15 such planes. (courtesy Christina Thaller.)

the atomic model of the protein (with a predetermined amino acid sequence). A portion of an electron density map with the experimental fit of an atomic model is shown in Figure 2-3. The model and map in this example are generated by a computer.

Not all macromolecules of interest crystallize or form fibers easily; thus x-ray diffraction is not always useful for structural analysis. Since use of the electron microscope does not, in principle, require ordered arrays of macromolecules, its potential uses are more numerous. Another difference between these two methods of examining molecular structure is their *limit of resolution*. Two objects are said to be *resolved* if they can be distinguished as two separate objects. The limit of resolution is the closest two objects may be to one another and still be resolved. The resolution of the electron microscope is low, approximately 20 Å, in comparison to x-ray diffraction methods, which are limited to about 1 Å resolution. The effects of a limiting resolution of 20 Å on an image of a protein are shown in Figure 2-4. Thus the use of the electron microscope is limited to low-resolution imaging of proteins, nucleic acids, viruses, and other large macromolecules (Figure 2-5).

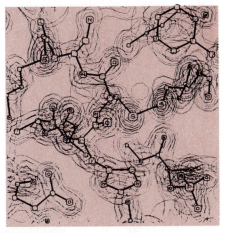

Figure 2-3
Computer-generated image of a superposition of an electron density map of trypsin with an atomic model showing the atom centers, van der Waals radii, and bonds. A single plane of the three-dimensional map and model is shown here. The resolution is 1.4 Å. Note that sulfur atoms have the highest electron density.

Figure 2-4
View of an uranyl acetate-stained synaptic edge of acetylcholine receptor membrane vesicle (**a**) and schematic representation of the 55 Å-long, funnel-shaped protrusions of the receptor molecules (**b**). (Reproduced from Kistler J, Stroud RM, Klymkowsky MW, et al: *Biophysical Journal* 1982; 37:371–393. By copyright permission of the Biophysical Society.)

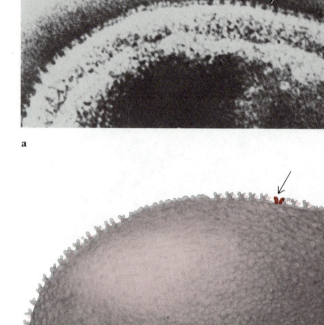

a

b

Common Structural Features of Proteins and Nucleic Acids

Proteins and nucleic acids, like all biological macromolecules, share a number of structural features. Both are large and complex, and each is formed by successive additions of repeating smaller units. Amino acids are linked end-to-end to form a protein molecule, and nucleotides are linked end-to-end to form a nucleic acid. These chains are properly described as being linear and unbranched. Both types of chains also have *polarity*, or direction. The left end of a protein sequence is customarily called the *N terminus*; the left end of a nucleic acid sequence is the *5' end*. The right end of a protein sequence is the *C terminus*; the right end of a nucleic acid sequence is the *3' end*. Figure 2-6 illustrates a linear segment of a protein and nucleic acid.

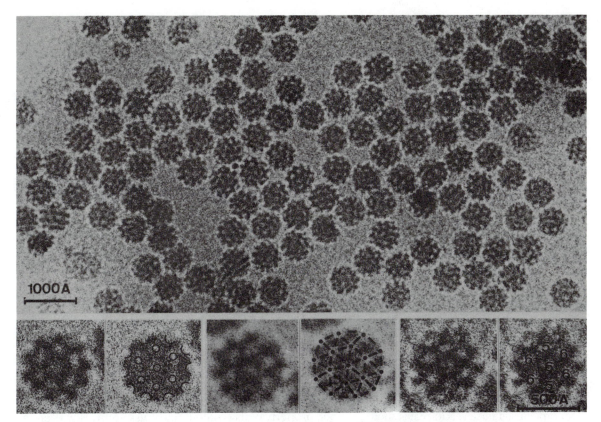

Figure 2-5
Electron micrographs of Semliki Forest virus. Individual virus particles are shown in the bottom panel. The axes of symmetry are indicated by circles (*left*), dots (*middle*), and numbers (*right*). The numbers indicate which symmetry axes are fivefold and sixfold. (From Adrian M, Dubochet J, McDowall AW: *Nature* 1984; 308:32–36. Reprinted by permission. © 1984 Macmillan Journals Ltd.)

Primary Structure

Primary structure is the sequence of the monomer amino acids or nucleotides. The primary structure of a macro-molecule can be written as a very long word (Figure 2-7). The arrangement of letters in the word somehow gives the molecule its specific properties, whether it be an enzyme or a nucleic acid. Functional portions of the molecule, as labelled in Figure 2-7, are not apparent in the primary sequence. These features are more readily seen in the folded three-dimensional structures.

Although both types of molecules appear complex even when written in a simple linear representation, their biological function arises not from the linear se-quence of their monomer components but from the three-dimensional conformation that they assume in water. The conformation a molecule assumes is an intrin-

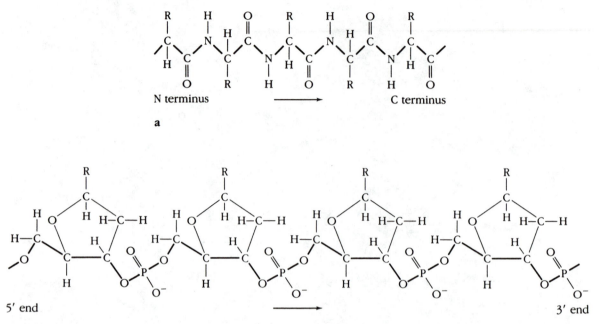

N terminus ⟶ C terminus

a

5′ end ⟶ 3′ end

b

Figure 2-6
Both protein and nucleic acid chains show polarity. **a**, The protein chain, by convention, begins at the N terminus and proceeds to the C terminus. **b**, The nucleic acid chain proceeds from the 5′ (position of the phosphate on the ribose) to the 3′ direction. Both chains are linear.

sic property of its covalent linear assembly. However, the precise final arrangement of atoms in three-dimensional space is determined by much weaker noncovalent forces, specifically the interaction of hydrophobic portions of a molecule and a complementary internal network of hydrogen bonds. Figure 2-8 shows the hydrophobic core that forms on folding a protein chain. For proteins, the pattern of nonpolar and polar amino acid side chains along the linear chain determines folding (see "Protein Folding" in Chapter 4). The hydrophobic portions associate in a hydrophobic core as the chain assumes the three-dimensional structure. The polar side chains point into the solution. In the case of nucleic acids, hydrophobic portions of the molecule (the bases) are stacked inside the surface formed by the extremely polar phosphate groups, which point into the solvent (see Figure 2-9 on p. 18). Thus the linear structures fold in complex and specific ways to form higher-order structures such as long rods, extended sheets, or globular molecules in solution.

M E Q R I T L K D Y A M R F G Q T K T A

K D L G V Y Q S A I N K A I H A G R K I
<u>DNA recognition helix</u>

F L T I N A D G S V Y A E E V K P F P S

a   N K K T T A

G C G G A U U U A G C U C A G U U G G G

A G A G C G C C A G A C U G A A G A U C
<u>Anticodon</u>

U G G A G G U C C U G U G U U C G A U C

b   C A C A G A A U U C G C A C C A
<u>Amino acid acceptor</u>

Figure 2-7
Linear sequences of a tRNA molecule (**a**) and the *cro* repressor (**b**). Relevant features of these structures are indicated but are not apparent from the linear arrays.

Secondary Structure

An attempt to understand the complicated structure of a polypeptide is greatly simplified by realizing that much of the complex three-dimensional (or tertiary) structure can be described as an assembly of regular *secondary structural* elements. Secondary structure is a regular spatial arrangement of the chain of nucleic acid or polypeptide. The α helix (pp. 64–67) and the β sheet (pp. 76–84) found in proteins are familiar secondary structural elements. In elements of secondary structure, the relationship between succeeding residues is repetitive and can be simply described. The spatial relationship between one residue and the next is described by a specific translational length and rotational angle. Thus a long segment of chain will fold according to a reiterated operation as illustrated in Figure 2-10. Generally, this involves rotation and translation of successive residues in the backbone. For example, in an α helix, each residue is rotated 100° relative to the previous one and is translated along the helix axis by about 1.5 Å. In other cases the symmetry is more complicated. For example, the collagen molecule is composed of three individual helical strands wound around each other to form a three-stranded superhelix (see "The Collagen Helix" in Chapter 5). DNA is another example in which rotation and

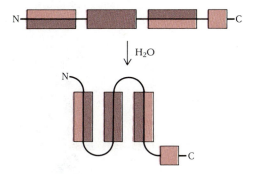

Figure 2-8
Folding of a protein chain in water. The schematic diagram shows (in two dimensions) the hydrophobic segments (shaded), which rearrange to form a compact core surrounded by the hydrophilic portions.

Figure 2-9
Computer-generated image of DNA in the B conformation. The van der Waals radii of the atoms are drawn. Oxygens are colored, nitrogens are color shaded, and carbons are gray. Only the atomic surface, which can be "touched" by a probe water molecule rolling over the van der Waals surface, is colored. The surfaces that are not accessible to the probe water molecule are left white. The shape of the white surface corresponds to the track left by the probe. The effect is to fill the solvent, inaccessible crevices of the van der Waals spheres with "grout." The resulting shaded surface is then the chemically active surface. (From Connolly M: *Science* 1983; 221:709–713. Copyright 1983 by the American Association for the Advancement of Science.)

translation of a repeating unit (in this case a pair of nucleotides) generates an element of secondary structure (in this case, a double helix). (See pp. 99–104 for a discussion of A- and B-DNA.)

Tertiary Structure

Tertiary structure is the complete three-dimensional structure of a macromolecule. The precise positions of individual atoms are specified in a tertiary structure, and this information can only be gotten from x-ray diffraction studies. For both proteins and nucleic acids the side chains cause the flexible backbone to assume a unique tertiary structure.

In later chapters, several examples of the precise tertiary structure of polypeptide and nucleic acid backbones will be described and discussed. The illustrations showing all-atom representations provide tertiary structural information for the side chains.

Symmetry ɅɹʇǝɯɯʎS

Symmetry is found everywhere in biology. It is necessary to understand the concepts of symmetry to see how it affects biological structures. Despite its apparent simplicity, seeing symmetry can be surprisingly difficult.

Intact individual backbone units from the model kit will be used to demonstrate some simple aspects of symmetry. You can construct the objects pictured in Figures 2-11 to 2-14 without gluing the parts.

An array of identical (or nearly identical) objects can be described by defining the object and its positions in space. If the designation of these positions follows simple rules, the array of objects possesses *symmetry*. This is a familiar notion in everyday life. Symmetry is observed daily, although often unnoticed, in floor tiles or patterns on fabric. Symmetry in the biosphere is also familiar, as in flower petals and in the bodies of creatures such as sea stars.

Recognizing the symmetry or pseudosymmetry that occurs in molecules simplifies their description and greatly aids understanding and analysis. This will be apparent in studying models of the structures in this book. For our purposes, we can define the necessary elements of symmetry simply.

Rotational Symmetry

Objects are related by rotational symmetry if two identical objects can be brought into spatial coincidence by a rotation of one of them about some axis (which must be determined by inspection). If the rotation is 180°, the pair possesses *twofold symmetry*. The term twofold comes from the calculation 360°/2 = 180°; "fold" means "times" in the sense that after two such rotations, the cycle of

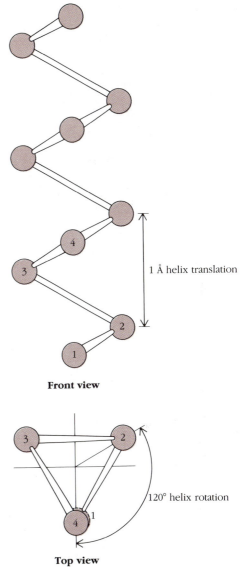

Front view

1 Å helix translation

120° helix rotation

Top view

Figure 2-10
Helices are the most common geometric forms assumed by long-chain polymers. This shows a simple example with a single unit that is repeated in the final structure by a 1 Å translation and a 120° rotation.

Figure 2-11
This object has one twofold axis of symmetry as shown. It is perpendicular to the page.

rotations is complete. Similarly, fourfold rotational symmetry would relate four identical objects by 360°/4 = 90° rotations. Figure 2-11 shows two objects (which have been joined into one larger object) related by twofold symmetry. Note the position of the axis of this symmetry in Figure 2-11. This imaginary axis is termed a *twofold axis*.

Twofold axes are common in biological structures. Less common are threefold, fourfold, fivefold, and sixfold axes. These are observed in the arrangement of protein subunits in certain complex assemblies. The protein coats of many spherical viruses are an arrangement of proteins related by twofold, threefold, and fivefold axes (see Figure 2-5). Even a seventeenfold axis has been observed in the helical tobacco mosaic virus. The seventeenfold symmetry is apparent in cross section through an image of a disk, a structure seen during the assembly of the virus (Figure 2-12).

Translational Symmetry

Translational symmetry is equally familiar. The cars in parking lots are usually related by translational symmetry. Translational symmetry can be observed in one, two or three dimensions. The direction of translation defines the symmetry axis, and the description is complete by specifying a fixed translation distance. Figure 2-13 shows an array of objects related by translational symmetry in two dimensions. Translational symmetry is not uncommon in biological structures (for example, in a bee's honeycomb), although often it is observed as part of coupled rotational-translational symmetry.

Helical Symmetry

A combination of these two kinds of symmetry generates a *helix*. The helix is a rotation of a fraction of 360° followed by a translation of the object by a fixed distance. The rotation and translation both take place about the same axis—the helix axis. Naturally, the complete helical

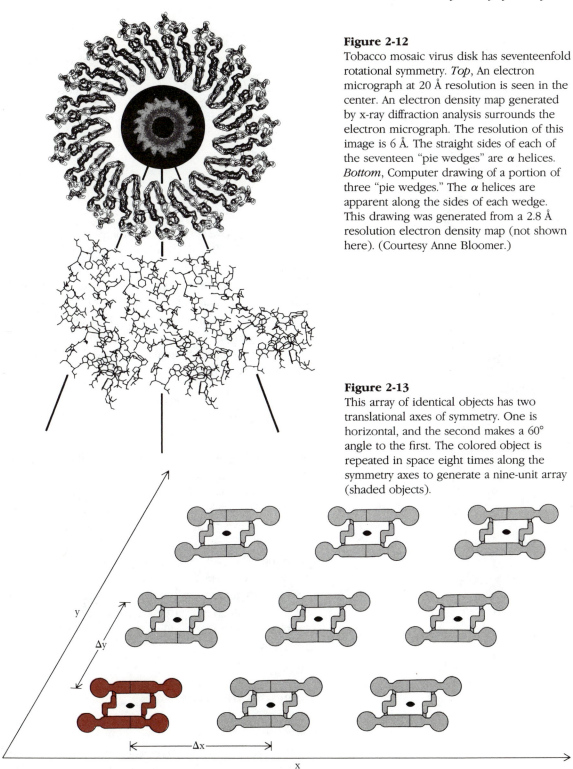

Figure 2-12
Tobacco mosaic virus disk has seventeenfold rotational symmetry. *Top*, An electron micrograph at 20 Å resolution is seen in the center. An electron density map generated by x-ray diffraction analysis surrounds the electron micrograph. The resolution of this image is 6 Å. The straight sides of each of the seventeen "pie wedges" are α helices. *Bottom*, Computer drawing of a portion of three "pie wedges." The α helices are apparent along the sides of each wedge. This drawing was generated from a 2.8 Å resolution electron density map (not shown here). (Courtesy Anne Bloomer.)

Figure 2-13
This array of identical objects has two translational axes of symmetry. One is horizontal, and the second makes a 60° angle to the first. The colored object is repeated in space eight times along the symmetry axes to generate a nine-unit array (shaded objects).

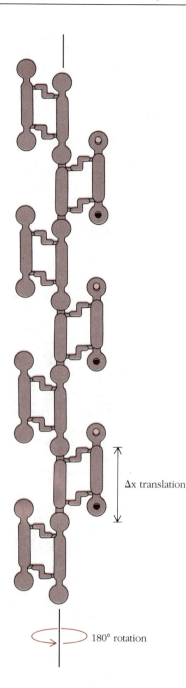

Δx translation

180° rotation

Figure 2-14
The arrangement shows translational and rotational symmetry coupled to give helical symmetry. The translational axis is straight and is along the long axis. The rotational component of the helical symmetry is 180° (a twofold rotation). There are no other symmetry elements. No single simple rotational operator will work on this helix.

structure may have rotational symmetry in addition to the rotational-translational symmetry that defines the helix. DNA is one such example.

Figure 2-14 shows an example of helical symmetry in which a twofold axis is along the translational axis. Note that this array does not have additional rotational symmetry aside from the helical symmetry, although the building blocks (pairs of model units) have local perpendicular twofold axes.

Summary

X-ray crystallography is used to determine the tertiary structures of proteins and nucleic acids. Both types of molecules are made of linear chains of repeating chemical units. The backbone structure has polarity. The side chains that are covalently linked to the backbone are positioned at precisely repeated intervals along this backbone. The pattern or sequence of side chains give the molecule its special functional properties.

Symmetry pervades biological structures. Translational and rotational symmetry combine to give helical symmetry—the most widely observed spatial symmetry in biology.

Suggested Readings

Diffraction Analysis

Blundell TL, Johnson LN: *Protein Crystallography,* New York, Academic Press, Inc., 1976. The only reference book available devoted exclusively to protein cystallography. This text covers all aspects of macromolecular structure determination in theory and practice and is intended for advanced students.

Stout GH, Jensen LH: *X-ray Structure Determination*. New York, MacMillan Company, 1968. Simple, practical guide to the methods and theory of x-ray crystallography as applied to structure determination for single crystals of small molecules. It is one of the simplest available descriptions of x-ray crystallographic methods.

Symmetry

Shupnikov AB, Kopsik VA: *Symmetry in Science and Art*. New York, Plenum Press, 1974. Rather long discussion of the symmetry principles found in art and in nature. Symmetry of molecular structures are discussed. The book is intended for advanced students.

DNA Sequencing

Gilbert W: DNA sequencing and gene structure. *Science* 1981; 214: 1305–1312. The Nobel Prize address of the author describing the development of the chemical methods for DNA sequencing and their use in determining gene structure.

Sanger F: Determination of nucleotide sequences in DNA. *Science* 1981; 214: 1205–1210. The Nobel Prize address of the author describing the development of enzymatic methodology for sequencing DNA.

Protein Sequencing

Laursen RA, Machleidt W: Solid phase methods in protein sequence analysis. *Methods Biochem Anal* 1980; 26: 201–284. Review of recent methods for protein sequence analysis. The article is written in great detail and is primarily intended for advanced students.

Virus Symmetry

Klug A: Architectural design of spherical viruses. *Nature* 1983; 303: 378–379. Discussion of the most recent aspects of the symmetry principles for the spherical viruses.

Butler PJ, Klug A: The assembly of a virus. *Sci Am* 1978; 239: 62–69. Very readable discussion about the assembly and symmetry principles for the tobacco mosaic virus.

Prediction of Protein Secondary Structure

Chou PY, Fasman GD: Prediction of the secondary structure of proteins from their amino acid sequence. *Adv Enzymol* 1978; 47: 45–148. Extensive review on the methods used for predicting secondary structure of a globular protein based on the amino acid sequence. It is easy to read, but the next reference should be consulted as an example of the serious pitfalls of secondary structure prediction.

Kabasch W, Sander C: On the use of sequence homologies to predict protein structure: identical pentapeptides can have completely different conformations. *Proc Natl Acad Sci USA* 1984; 81: 1075–1078. Discussion about how the short peptides of at least five amino acids can assume a variety of secondary structures from α helix to extended β strands.

Electron Microscopy

Tinoco I, Sauer K, Wang JC: *Physical Chemistry*. Englewood Cliffs, NJ, Prentice-Hall, Inc., 1978, pp. 581-590. Good basic introduction to electron microscopy.

3

How to Build the Models

Building the backbone models is easy and fun if you follow the instructions carefully. The two most important points are to be sure to check the angle settings, and to allow the glued model parts to dry fully. It is not necessary to read this chapter before the rest of the text, but you should be sure to read it carefully before you attempt to construct any models.

Before starting to construct a model you should locate a clean flat surface with good, bright lighting. You should also have available a marking pen, 15 cm ruler, pair of pliers for cutting the metal support rods, and the model kit. Throughout the text tables listing the angles that are used to build the models are included. We describe in this chapter how to use the angles to build the models.

The Model Kit

Scale

The scale of the model parts for building proteins is 100 mm = 1.0 nm. This corresponds to a magnification of 10^{-1} m/10^{-9} m = 10^8. In the text, centimeter (cm; 10^{-2} m) and angstrom (Å; 10^{-10} m) units are used because they are more convenient. The scale of the protein models is 1 cm/Å.

Figure 3-1
Individual backbone units as they appear in the Blackwell kit.

The scale of the nucleic acid models is approximately two-thirds that of the proteins. This is because the typical distance between α-carbons is 3.8 Å, whereas the typical phosphate-to-phosphate distance is 6.3 Å.

Backbone Units

Backbone units are the fundamental building blocks of *Blackwell Molecular Models* (Figure 3-1). The model kit contains backbone units in two different colors. Each

Figure 3-2
An individual backbone unit consists of three parts. A disposable plastic runner connects parts V and S.

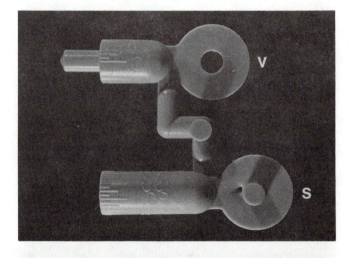

Figure 3-3
Part S (*top*) has markings indicating the angle scales. Part V (*bottom*) has markings for the vernier scale.

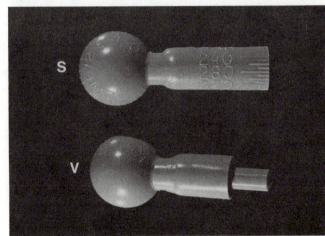

backbone unit consists of three pieces (Figure 3-2). A runner, which is disposable, connects the two other pieces *V* and *S* (Figure 3-3). Part V has a hole in the center of the hemisphere (Figure 3-4), and part S has the rod in the hemisphere (Figure 3-5). The hemispheres snap together to form a ball and the completed backbone unit (Figure 3-6). Part V becomes the N terminus or 5′ end of the unit, and part S becomes the C terminus or 3′ end of the unit.

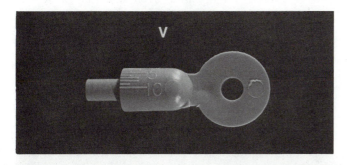

Figure 3-4
Part V has a hole in the hemisphere.

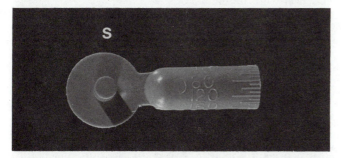

Figure 3-5
Part S has a rod protruding from the hemisphere.

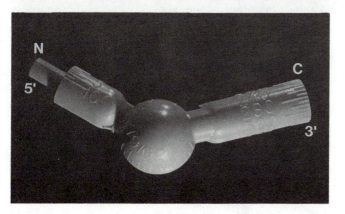

Figure 3-6
Assembled backbone unit. Note that the two halves are the same color. Part V (*left*) is the N terminus or 5′ end of the unit. Part S (*right*) is the C terminus or 3′ end.

Figure 3-7
Model glue.

Model Glue

Plastic cement is used to permanently fix the θ and τ angles used in the construction of the models (Figure 3-7). As in building any plastic model, adequate glue must be used to ensure adhesion of the parts, and the glued parts must be allowed to dry thoroughly before they are manipulated further. Remember that the glue is an eye irritant and must be used with adequate ventilation and eye protection. Acetone or nail polish remover may be used to remove glue from your hands. Do not attempt to unglue model parts using acetone; it is a solvent that also dissolves the models.

Connectors

The connectors are clear plastic pieces shaped like a cup (Figure 3-8). The cup fits around the completed backbone unit (Figure 3-9). The connectors are used in conjunction with the metal rods.

Rods

The rods are used to fix distances between specific backbone units in the completed model (Figure 3-10*a*). Rods are cut with a standard wire cutter (side-cutting or conventional pliers) to the lengths given in the building instructions in this text. Two connectors are attached to the ends of the rods (Figure 3-10*b*). This rod assembly is then used to connect pairs of backbone units in the models, as shown in Figure 3-11).

Setting the Angles

The Vernier Scale

Models are constructed by attaching the model parts and setting the two angles θ and τ. The θ angle is the *bend angle* which relates the two hemispheres of the ball in

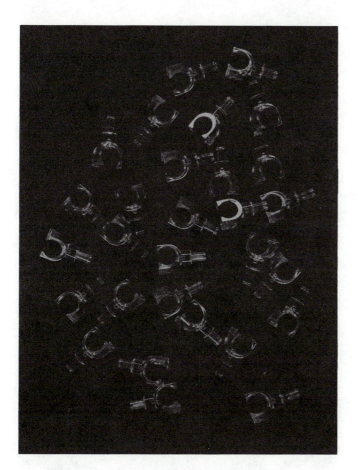

Figure 3-8
These connectors are used with the metal rods (Figure 3-10) to connect and support pairs of backbone units in the completed model.

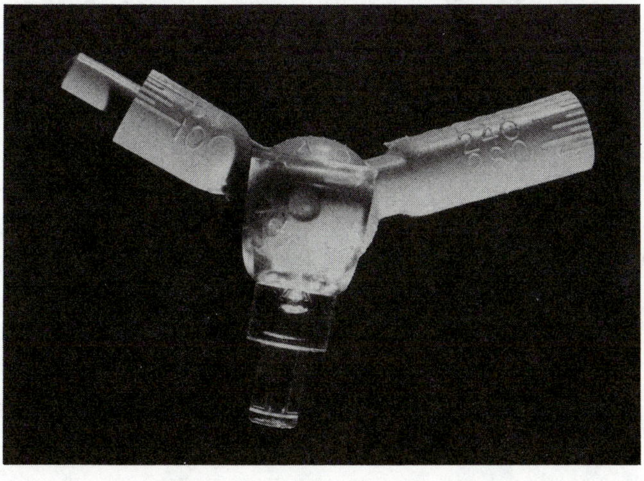

Figure 3-9
The clear plastic connector snaps over the ball of the complete backbone unit.

Figure 3-10
a, The rods are inserted into two connectors to form the supports. The metal rods are cut to the lengths indicated in the tables for construction of each model. **b**, Completed rod assembly. The distance measured from the centers of the two connectors is 2 cm longer than the metal rod.

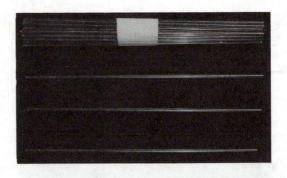

Figure 3-11
Pairs of backbone units are connected by completed rod assemblies. This is a close-up view of the support rods in the collagen model.

each backbone unit. The τ angle is the *torsional angle* between adjacent backbone units. For all the models presented in this book, angle settings are listed in tables. Both θ and τ can be set precisely to the degree using a vernier scale.

To read the vernier, you need bright light. Look at part S. On the hemisphere are angle scale markings. These are numbered 0, 4, 8, and so on and correspond to the bend angle (θ) of 0°, 40°, and 80°, and so on. The markings on the shaft are numbered from 0 to 320, corresponding to the torsional angle (τ) of 0° to 320°. These are the *angle scales*. Part V has two *vernier scales,* one on the hemisphere used to set the angle θ, and one on the shaft used to set the angle τ. The vernier scale has markings from 0° to 10°. These allow fine adjustment of angle settings for the unit degrees (1° to 9°, 11° to 19°, and so on).

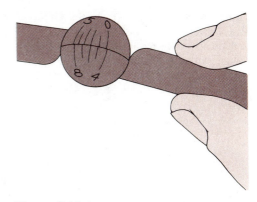

Figure 3-12
The θ angle on this completed backbone unit is set at 42°. The 0 on the vernier scale is set near 40 on the angle scale, and the 2 marking on the vernier aligns with the 60 line on the angle scale.

Reading the Vernier

The vernier is read by aligning the 0° line of the vernier scale with the desired 10° line of the angle scale (that is, 10, 20, 30, and so on). This sets the angle crudely to 10°, 20°, or 30°, or more. The *unit angle* is set by rotating the vernier scale slightly counterclockwise to align the appropriate unit degree marking on the vernier with any line on the angle scale. If you look at the angle scale now the 0° line on the vernier points between two 10° markings on the angle scale. The value of 42° is set for the angle θ on the ball, as shown in Figure 3-12. Figure 3-13 shows the torsional angles as set on the rod to be 124°.

Practice setting the vernier using one or two backbone units (do not use the glue). Try setting different angles until you feel comfortable with the use of the vernier scale.

Two features of the individual units should be noted. The angle θ is measured between an imaginary line extending from the N or 5′ side (part V, containing the vernier) and the C or 3′ side (part S, containing the angle scale). Thus the actual angle you *see* between the two halves of the unit is $180° - \theta$ (Figure 3-14). Do not be

Figure 3-13
The τ angle is set at 124°. The 0 on the V unit points between 120 and 130 on the S unit, and the 4 marking on the V unit is aligned with the 160 line on the S unit.

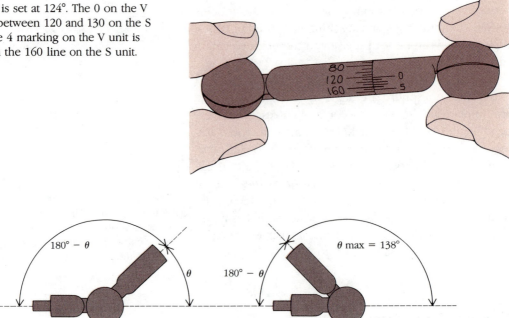

Figure 3-14
a, The θ angle is 180° minus the angle that the completed backbone unit makes. **b**, The maximum θ angle is 138°. The angle you see between the two halves of the backbone unit is 32°.

confused by this. A consequence of this is that the largest possible θ angle is approximately 140°. The units will not rotate any further. Of course the τ angles can be anywhere between 0° and 360°. The only limitations are the structural constraints of the molecule you are constructing.

The vernier scales allow angles to be set with an accuracy of 1°, although this precision is not required for building accurate models. Random errors of up to ±5° will not seriously alter the final structure of the model. This is because connecting rods, which are added at the end of construction, serve to fix the model in its proper conformation.

Building a Model

Step 1 On a sheet of 8½ × 11 inch paper, write the numbers 1 to 10 in a row lengthwise on the paper. Below this row of numbers, write the numbers 11 to 20 in the same fashion. The next row should contain the numbers 21 to 30 and the last row, 31 to 40. You will lay the glued backbone units down in order on this paper (Figure 3-15).

Step 2 Locate the θ angle in the table of angle settings for the model you intend to build. Place a part V in one hand. Find a part S and lay it nearby. Brush some model glue into the depression and onto the flat surface of the ball of part V. Remember to use enough glue, but not so much that it squirts out when the two hemispheres are squeezed together. Snap the part S hemisphere onto the part V hemisphere. Squeeze the ball halves together tightly so there is no gap. Using the vernier, set the bend angle θ to the value given in the table for that unit.

Step 3 Lay this unit on your numbered sheet in position 1. Prepare ten successive backbone units as in step 2. Lay these in a linear array on your sheet. Repeat this for all backbone units. The units should be arranged on your numbered sheet(s) in order. Do not mix them up! Allow the backbone units to dry thoroughly before proceeding to set the torsional angles τ. The units take at least 15 to 30 minutes to dry to the touch and 2 hours to overnight to dry completely. *It is a good idea to have someone else check the settings for the bend angles before the glue has set.* Model building may be continued after the pieces have dried to the touch. You may test them by trying to change the angles using a small amount of force. If you force them too much, they will come apart, so you should be careful when setting the torsional angles.

Step 4 Torsional angles will be set for segments of ten backbone units (or greater or fewer units as mentioned in the tables for certain models). Find the τ angles in the table for the first two units. Apply glue generously in the hole of backbone unit 1 and on the flat cross section of the shaft. Insert backbone unit 2 into unit 1 and set the τ angle using the vernier scale. Note that the τ angle is set at the C or 3' side of a backbone unit using the vernier on the next unit. This is illustrated in Figure 3-16.

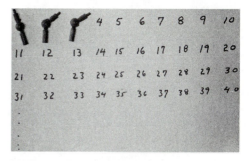

Figure 3-15
The completed backbone units should be arranged in order corresponding to the instructions.

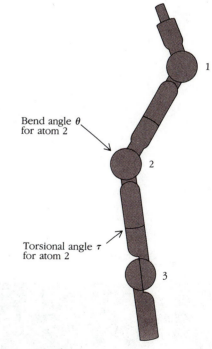

Figure 3-16
The τ angle between units n and n + 1 is set using the vernier scale on the next unit as shown here.

Step 5 To set the next τ angle you must proceed *carefully*. Do not touch backbone unit 1. Hold backbone unit 2 carefully, apply glue, and insert unit 3. Proceed carefully to finish the ten-unit segment. When you have finished, recheck the angles to make sure you have not changed any of them inadvertently. *It is a good idea to have someone else check the settings for the torsional angles before the glue has set.* Lay the completed ten-unit segment aside to dry thoroughly. You should label every tenth unit 10, 20, and so on, using a permanent magic marker (Sharpie, or VWR Lab Marker), small adhesive labels or tape. These numbers will aid further model construction.

Assemble the next segment with the τ angle for unit 11. Construct all the segments and set them aside. You should note that the last τ value is *not* used to set any torsional angle.

Step 6 The fully glued ten-unit segments are assembled into the final structure by gluing the τ angles between units 10 and 11, 20 and 21, and so on using the τ angles for units 10, 20, and so on.

Step 7 Support rods are added to bring the backbone into the correct conformation and to stabilize the finished model. Cut the support rods to the lengths given in the instructions for the model you are building. These rods are 2 cm shorter than the actual distance in the model. Put connectors on both ends of the rod, pushing the connectors onto the rods firmly and completely. The connectors add 2 cm to the length of the metal rod. The completed rod assembly is now snapped onto the ball portions of the units given in the model instructions. You will need to turn the connectors on the rod to seat them properly on the balls. You will also notice that the connector cup halves are two different sizes so that they will fit on the balls. You may need to trim (with scissors or side-cutting pliers) the smaller connector cup halves to fit onto units with very large θ values (that is, with very acute angles). For Z-DNA (pp. 104–108) and tRNA (pp. 109–115) you will need to cut the small cup half off altogether as described. Figure 3-17 shows this special completed rod assembly. Once all the support rods are

Figure 3-17
This special rod assembly is used in the construction of models for tRNA and Z-DNA. The small clips have been cut off the connectors.

in place and the model structure has been checked against figures in the text, the connectors may be glued to the backbone units. Place a drop of glue on the ball at the edge of the connector. It will flow into the space between the connector and ball. Once the glue has completely dried, the model may be handled freely. Congratulations! You have finished building your first backbone model.

Repairing Mistakes

You may be able to repair incorrect settings after the glue has set. Find the error in your model. It is easiest to correct errors in bend angle settings. Use a sharp knife (x-acto knife or scalpel with a pointed blade). Insert it between the two halves of the ball and work your way around the ball. You may be able to separate the two ball halves with the rod intact. If not, cut through the rod, leaving it in the cavity of the other hemisphere. Apply more glue, and reset the angle. If the markings have been dissolved by the glue, set another unglued model part to the appropriate setting and use it as a template. Torsional angles may be corrected the same way, although it is more difficult. You may wish to cut the bend angles on either side of the incorrect torsional angle and simply replace the entire unit.

After you have made the correction, you must allow the parts to dry thoroughly, possibly overnight. Because the structural integrity of these parts is disturbed, they are more fragile than the rest of the model.

Summary

To build a model, you should find a clean place and assemble all the materials. There are four steps in building a model. The bend angles, θ, are set to the angles found in the construction tables and glued. Torsional angles are set and glued to form chain segments that are ten units long. The segments are joined and glued. Finally, support rods are added to link specific points in the backbone.

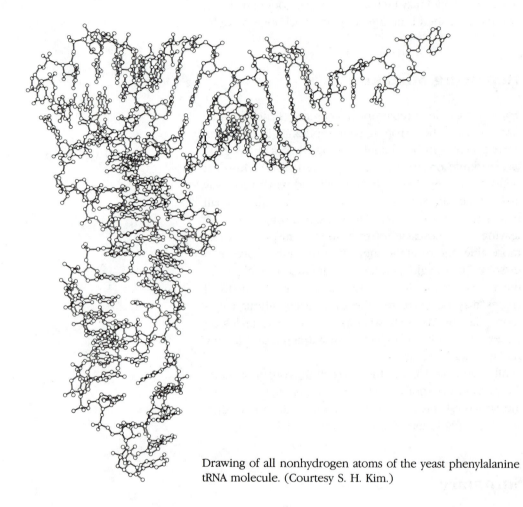

Drawing of all nonhydrogen atoms of the yeast phenylalanine tRNA molecule. (Courtesy S. H. Kim.)

PART II

PROTEIN AND NUCLEIC ACID STRUCTURES

Architectural principles govern the shape and function of structures. The principles of efficient close-packing, symmetry, and modular construction are observed in the three-dimensional structures of proteins and nucleic acids. The simple beauty of the DNA double helix is transformed to elegant complexity in the tRNA molecule. Complex protein structures are made of linked components of regular structural segments, each of which is simple and elegant in design.

4

Protein Structure

Protein structures are complex. To understand them the nature of the backbone and side chains must be examined in detail. The rigid peptide link has the property of forming two hydrogen bonds, one as donor and one as acceptor. Each of the links in the backbone is free to rotate into positions that are allowed by the side chains. A few of these rotations give rise to orientations of the backbone that allow all donor and acceptor hydrogen bonds to be made. These are the regular secondary structures observed for the α and 3_{10} helix, collagen, and β sheets.

The Polypeptide Backbone

The repeat unit of the *main chain* is composed of the six atoms N, H, Cα, H, C$'$, and O. Three atoms repeat in the polypeptide backbone, the amino nitrogen (N), the α-carbon (Cα), and the carbon atom of the carbonyl group (C$'$). The α-carbon atoms bear the side chains that differentiate the 20 amino acids. Individual amino acids are linked into polypeptides by the peptide bond. As Figure 4-1 shows, the peptide bond is formed between the carbonyl carbon of one amino acid and the amino

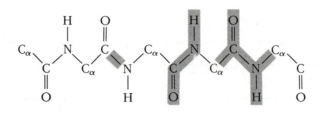

Figure 4-1

Amino acids are joined by peptide bonds. The backbone atoms are arranged in the order N, Cα, and C′ from N to C terminus (left to right). The peptide bond (highlighted left) is the amide linkage between CO and NH, the peptide group (highlighted center) is the CO—NH unit, and the peptide link (highlighted right) is the entire assembly of four atoms and six bonds that links α-carbons. Note that only two atoms and three bonds of the peptide link actually serve to connect adjacent carbons.

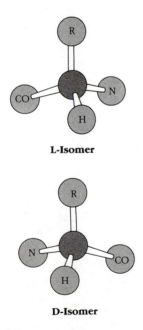

L-Isomer

D-Isomer

Figure 4-2

L and D amino acids. In the naturally occurring L amino acids, the atoms CO, R, and N are read clockwise as one looks down the H → Cα bond. CO, R, N, and H atoms are indicated.

group of the next amino acid. The four atoms—the amino nitrogen and hydrogen, the carbonyl carbon and oxygen—are referred to as the *peptide group*. This structure is also called the *peptide link*, since it links the α-carbons in the polypeptide backbone. The peptide link is a rigid and well-defined structure. Thus a polypeptide can be simplistically described as a set of α-carbons that are attached by intervening peptide links to form the backbone.

Chirality

The α-carbon atom in the amino acid makes four bonds with four different atoms. Thus all amino acids (with the exception of glycine) must be described as having a *handed* or *chiral* nature. The three-dimensional structure of a protein reflects the chirality of the α-carbon atom as defined in Figure 4-2.

Because secondary structure is a result of the regular folding of the polypeptide backbone, secondary structure elements often have a handedness themselves. The definition of handedness, or chirality, is different from that used for the amino acid α-carbon. The structures considered here are right- or left-handed by the same criteria as applied to the threads of a screw. Turning a right-handed screw clockwise causes it to advance. To determine the handedness of a three-dimensional structure, abstract the structural backbone to a coiled line in space. Then, imagine following the line. If you advance down the line in a clockwise direction, the structure is right-handed. Compare the two helices in Figure 4-3. Note that the handedness cannot change as you rotate the object or change your viewing direction. The handedness is reversed by viewing the object in a mirror. Try these simple experiments when you construct the α helix (p. 66). All

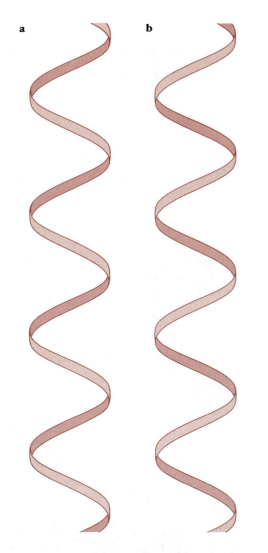

a   b

Figure 4-3
Left- and right-handed helices. Helix **a** is left-handed, and helix **b** is right-handed. Note that the handedness may be determined using the rules described in the text.

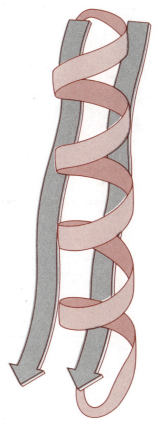

α helices are right-handed. Similarly, the three-stranded collagen helix is always right-handed, but each of its individual chains is a left-handed helix (p. 75). Even the assembly of secondary structural elements into tertiary structure can show handedness (Figure 4-4).

Amino Acid Side Chains

For water-soluble proteins, amino acid side chains may be simply divided into two groups—those that are found inside the protein and form the hydrophobic core and

Figure 4-4
This $\beta\alpha\beta$ unit is a common feature of many protein structures. The three secondary structural elements follow a right-handed path as shown in the diagram.

Table 4-1 Amino acid side chains and properties

Codes	Name	Formula	Comments	Where Found
G Gly	Glycine	—H	Allows flexibility and kinking of chain	Surface or buried; turns
A Ala	Alanine	—CH$_3$	Abundant	Surface or buried
V Val	Valine	—CH with CH$_3$ and CH$_3$	Stiff and bulky; major constituent of hydrophobic core	60%–90% buried
L Leu	Leucine	—CH$_2$—CH with CH$_3$ and CH$_3$	Stiff and bulky; major constituent of hydrophobic core	60%–90% buried
I Ile	Isoleucine	CH$_3$ over —CH—CH$_2$—CH$_3$	β-carbon is chiral; stiff and bulky; major constituent of hydrophobic core	60%–90% buried
F Phe	Phenylalanine	—CH$_2$—C (ring: CH=CH, CH, CH—CH)	Very hydrophobic; aromatic side chains require CH$_2$ spacer	50%–70% buried
Y Tyr	Tyrosine	—CH$_2$—C (ring: CH=CH, C—OH, CH—CH)	Only weakly acidic; H forms hydrogen bonds; can be phosphorylated	Same as Phe but more frequently at surface and often at turns

those that are on the surface and interact with the solvent. More than two of three side chains that compose the hydrophobic core are members of the set Cys, Val, Ile, Leu, Met, Trp and Phe (see Table 4-1 for full names of side chains). The amino acids that are usually found on the surface are Lys, Arg, His, Asp, Glu, Asn, and Gln. A few amino acids are commonly found in both the interiors and on the surfaces of proteins. These are Pro, Gly, Ser,

Table 4-1 (continued)

Codes	Name	Formula	Comments	Where Found
W Trp	Tryptophan	$-CH_2-C$ (indole ring: CH—NH, C—CH, C, HC=CH, CH)	Rarest and largest; bonds only as donor; difficult to pack	Often buried but commonly at surface as well
M Met	Methionine	$-CH_2^- -CH_2^- -CH_2^- -S-CH_3$	Very large; weak dipole	60%–90% buried
C Cys	Cysteine	$-CH_2^- -SH$	Forms disulfide bridges most commonly in extracellular proteins; can be catalytically active	60%–90% buried
S Ser	Serine	$-CH_2^- -OH$	Hydrogen bonds; catalytically active as its hydrogen atom can be removed; often derivatized (with phosphate, for example)	Surface or buried; turns
T Thr	Threonine	$-\overset{OH}{\underset{H}{C}}-CH_3$	β-carbon is chiral; hydrogen bonds; often derivatized, for example, with phosphate or sugars; S and T buried by forming hydrogen bonds with other hydrogen-bonding groups (usually on backbone)	Surface or buried
H His	Histidine	$-CH$ (imidazole ring: H$^+$, N—CH, N—H, CH, H)	Partial positive charge at pH 7	30% buried, often at active sites.

Table 4-1 (continued)

Codes	Name	Formula	Comments	Where Found
K Lys	Lysine	$-CH_2^- - CH_2^- - CH_2^- - CH_2^- - NH_3^+$	Positive charge at pH 7; floppy; makes proteins soluble; less frequently catalytically active; CH_2 groups make side chain partly hydrophobic	Surface
R Arg	Arginine	$-CH_2 - CH_2 - CH_2 - NH - C \begin{smallmatrix} \nearrow NH_2^+ \\ \searrow NH_2 \end{smallmatrix}$	Positive; often binds phosphates; very large; when buried, it is paired with E or D	Surface
D Asp	Aspartate	$-CH_2 - \overset{\overset{\textstyle O}{\|}}{C} - O^-$	Negative; catalytically active; less flexible than E	Surface
E Glu	Glutamate	$-CH_2 - CH_2 - \overset{\overset{\textstyle O}{\|}}{C} - O^-$	Negative; flexible; catalytically active	Surface
N Asn	Asparagine	$-CH_2 - \overset{\overset{\textstyle O}{\|}}{C} - NH_2$	Hydrogen bonds; uncharged; glycosylation site	Surface
Q Gln	Glutamine	$-CH_2 - CH_2 - \overset{\overset{\textstyle O}{\|}}{C} - NH_2$	Hydrogen bonds; uncharged	Surface
P Pro	Proline	$\begin{smallmatrix} (C) - CH_2 \\ \quad\quad CH_2 \\ (N) - CH_2 \end{smallmatrix}$	Important for bends in protein folding	20%–30% buried

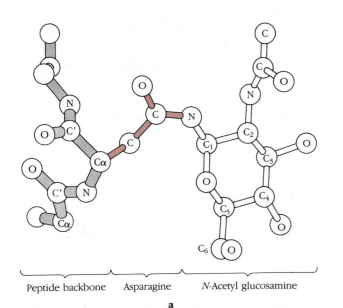

Peptide backbone — Asparagine — *N*-Acetyl glucosamine

a

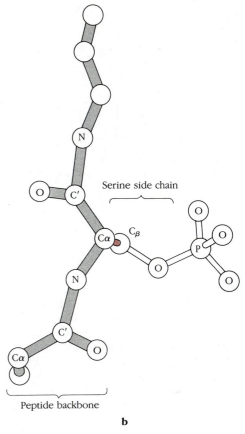

Serine side chain

Peptide backbone

b

Figure 4-5
Examples of posttranslational modifications.
a, The nitrogen atom of asparagine is covalently bonded to the C1 of *N*-acetyl glucosamine. **b**, The phosphorus atom is linked via an oxygen to the β carbon of serine. This is serine 14 of the enzyme glycogen phosphorylase.

Thr, and Tyr. Pro and Gly are hydrophobic and can be buried inside proteins (see models for cro repressor, insulin, and lysozyme). However, they are often found in turns which are usually located on the surface. Ser, Thr, and Tyr may be hydrophobic and buried inside the protein and not interact with water if the polar hydroxyl group is complemented by the formation of a buried hydrogen bond. This usually involves other side chains or backbone carbonyl groups.

The side chains of some amino acids function in the active sites of enzymes (Cys, Lys, Ser, Arg, His, Asp, Glu); the discussion of lysozyme (p. 181) includes an example of this. Some side chains may be covalently modified by the attachment of a phosphate group or sugar residue. (Ser, Thr, Tyr, Asn; Figure 4-5 illustrates a modified Asn and Ser). A portion of an antibody (immunoglobulin) molecule with a sugar group is shown in Figure 4-6. The properties and structure of the amino acid side chains are summarized in Table 4-1. Both the one- and three-letter codes listed in Table 4-1 are used in this book.

It should be noted that our division of amino acids into two groups (hydrophobic and hydrophilic), which are assigned two colors, is an oversimplification. The hydrophobicities of amino acids fall on a continuous scale, and some side chains show characteristics of both groups.

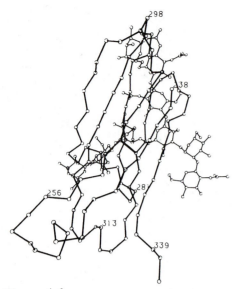

Figure 4-6

This α-carbon representation of a portion of the Fc region of an immunoglobulin (antibody) molecule shows the attachment site of an *N*-linked complex oligosaccharide. The carbohydrate is on the protein's surface. (Reprinted with permission from J. Deisenhofer, *Biochemistry* 20, 1981, pp. 2361–2370. © 1981 American Chemical Society.)

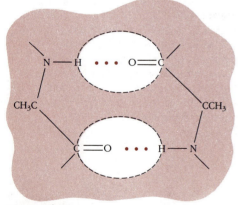

Figure 4-7

The hydrophobic core may contain the polar N–H and CO groups, but they must be paired by a hydrogen bond. The shading represents the hydrophobic core.

For example, the side chain amino group of lysine is positively charged at pH 7 and therefore should be solvated. However, the lysine side chain also contains four highly nonpolar methylene groups that are hydrophobic. Our classification is adequate for the purposes of model building. You will see that, in general, protein model surfaces are red, and the hydrophobic cores are green (see p. 174 on lysozyme structure).

Protein Folding

A protein molecule assumes its final three-dimensional conformation as a result of the chemical properties of the polypeptide backbone as well as the amino acid side chains. The protein backbone is a continuous filament of potential hydrogen bond donors (the amido hydrogens) and acceptors (the carbonyl oxygens). An important characteristic of protein molecules is that they are very compact. The interior of each soluble protein (in contrast to membrane proteins) is formed from tightly packed clusters of hydrophobic amino acid side chains (recall Figure 2-8). Within this hydrophobic matrix the number of hydrogen bonds formed between the peptide amido hydrogens or carbonyl oxygen groups is maximized. This is shown schematically in Figure 4-7. Noncomplemented amido hydrogens or carbonyl oxygens on the polypeptide backbone are rarely found in the hydrophobic interior of folded proteins. In contrast, the exterior of a protein will often have unpaired potential hydrogen bonding groups in the peptide backbone. Also, many of the side chains that are capable of forming hydrogen bonds, such as asparagine, glutamine and the charged amino acids, will be found freely solvated by water and not interacting with other side chains.

Even though proteins fold into compact structures, they assume a variety of different shapes, from filamentous to nearly spherical. The protein's primary sequence determines the shape, although exactly how this happens remains a mystery. Some physical principles that determine the shape of proteins are discussed next.

Why Proteins Fold

A description of the thermodynamics of protein folding in water is lengthy and complicated. Only the two most important interactions involved are summarized here.

Hydrogen bonds A noncovalent attraction forms among three atoms when two electronegative atoms, N and O, approach each other (Figure 4-8). One of the electronegative atoms must be covalently bonded to a hydrogen atom. The hydrogen bond forms by charge-charge attraction between the partially positive hydrogen atom and its electronegative partner as seen in Figure 4-8*a*. Hydrogen bonds are *linear*. In a nonlinear array of three partial charges, the electronegative charges will interact and repel each other (Figure 4-8*b*). Hydrogen bonds have a well-defined length. The distance between two hydrogen-bonded electronegative atoms is 2.8 Å (±0.2 Å). The hydrogen bonds between the functional groups (whether charged or uncharged) of the protein should not be any stronger than those between the functional groups and water, unless the protein-protein hydrogen bonds can be shielded from water, which dissipates the polar attractive force. Thus buried hydrogen bonds in the hydrophobic core (as in Figure 4-7) or partially buried hydrogen bonds lying on the protein surface contribute to the stability of the folded structure (Figure 4-9).

Hydrophobic Interaction These are not usually described as formal forces, but are a consequence of the forces between molecules of water. The hydrophobic groups associate with each other and not with water because water would rather self-associate than form weak attractive forces (van der Waals) with the hydrophobic groups. In effect, when water self-associates, it squeezes the hydrophobic groups out of solution and into clusters shielded from water by the more polar side chains. This is depicted in Figure 4-10. Thus proteins in water fold so that hydrophobic side chains are buried in

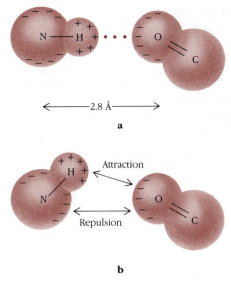

Figure 4-8
a, Schematic diagram showing the charge distribution in a hydrogen bond between an amino group and a carbonyl group. The *linear* array of charges on N–H··O is favored. **b**, The repulsion that would arise from a nonlinear array of the N–H and O atoms makes this arrangement less stable.

Figure 4-9
a, Hydrogen bonds are strong if water is excluded. **b**, A hydrogen bond near the surface of a protein is protected from being broken by water molecules in the solvent. **c**, Hydrogen bonds between NH and CO are weakened or broken if water molecules can compete for the hydrogen-bonding capability.

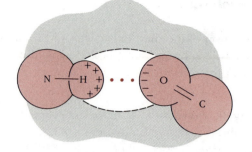

a

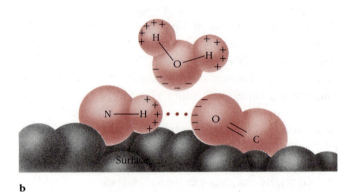

b

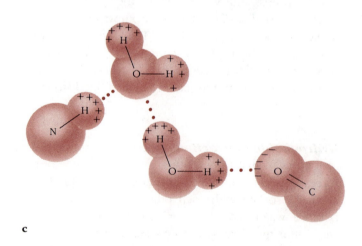

c

the interior and hydrophilic side chains are on the outside. Proteins in other solvents such as alcohols or in a vacuum will not demonstrate this feature and usually are not folded into the *native structure* (the stable form that is found in water).

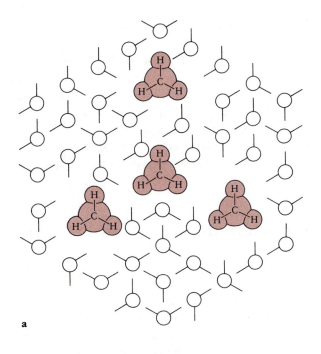

a

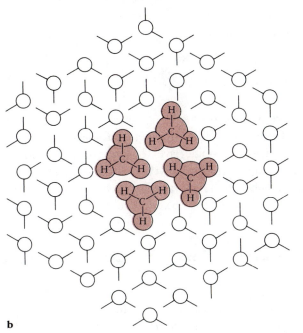

b

Figure 4-10
Schematic diagram of hydrophobic effect.
The water molecules compete for space
with the dissociated hydrocarbons (**a**). When
the hydrocarbons are dissociated, they have
more surface area, and there are fewer
interactions between the water molecules
(**a**) than when the four hydrocarbon groups
associate (**b**). Water prefers to interact with
itself, and in forming these interactions, the
hydrocarbons are forced to associate as
shown in **b**.

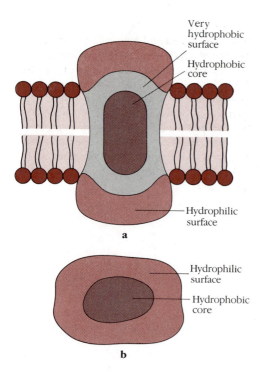

Very hydrophobic surface

Hydrophobic core

Hydrophilic surface

a

Hydrophilic surface

Hydrophobic core

b

Figure 4-11
The membrane protein shown schematically is inside out as compared to water soluble globular proteins. **a**, Membrane protein. In a cross section through the membrane, the hydrophobic side chains are directed outward into the lipid of the membrane, while the polar side chains point into the core of the protein, forming hydrogen bonds and a close-packed structure. The membrane and the extremely hydrophobic region is gray, the moderately hydrophobic core is color shaded, and the hydrophilic extracellular and cytoplasmic surfaces are colored. **b**, Soluble protein. In a cross section through the protein, the moderately hydrophobic core is evident. It is surrounded by the hydrophilic surface (colored).

The Role of van der Waals Forces

Van der Waals interactions are a less important contributor to protein structure. These weak attractive forces arise from the partial polarization of electrons in each atom that is induced when two atoms approach each other. Van der Waals forces do not determine protein structure because they occur between *all* atoms, including protein and solvent, and are not unique to the folded protein.

Characteristics of Folded Proteins

Membrane Proteins

Proteins in membranes, which are hydrophobic, are inside out (Figure 4-11). Their hydrophobic side chains are at the surface facing the hydrocarbons of the lipids, and the polar side chains interact with each other to form hydrogen bonds.

Bound Water

There is substantial evidence from x-ray crystallographic and spectroscopic techniques that soluble proteins are coated with a single layer of tenaciously bound water. This water can be removed only under extreme conditions of heat and high vacuum. The water shell is typically 30% of the solvated protein volume.

Domains and Subunits: Close-Packing

Nearly all proteins are composed of smaller structural units or domains (Figure 4-12). In some cases a protein is composed of a set of individual polypeptide chains, called *subunits,* that associate primarily by noncovalent interactions. These subunits may still fit together very

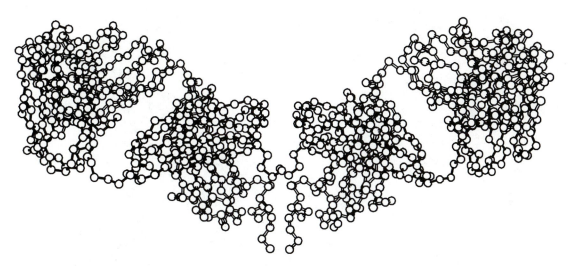

Figure 4-12
An immunoglobulin (antibody) molecule is made of four separate chains and folds into 12 domains of homologous tertiary structure. This figure shows an α-carbon representation of the structure of eight of these domains. Note the structural homology of the antiparallel β sheet domains.

tightly (Figure 4-13), much like the close-packed interior of a protein. The term *close-packed* means that the approximately spherical atoms fill space as efficiently as is theoretically possible (Figure 4-14). No more spherical atoms can fit inside a close-packed structure. Other proteins consist of a single, continuous polypeptide chain that folds into compact units known as *domains*. Because scientists cannot agree on the definition of the term domain, our definition is not meant to be rigorous but is simply to be used for the purpose of discussion. Domains are clusters of secondary structure elements that have associated into stable compact assemblies. A domain is composed of a number of segments of secondary structure which are linked by loops or turns (Figures 4-15 to 4-18).

A domain may be composed of one of the following:

- All α helices, as in one of the domains of hexokinase (Figure 4-15), or all of hemoglobin β subunit (Figure 4-16).

- All β strands that are antiparallel, as in one half of chymotrypsin (Figure 4-17).

- Mixed α and mostly parallel β structure, as in the nucleotide-binding domain of lactate dehydrogenase (Figure 4-18).

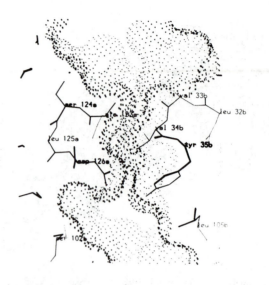

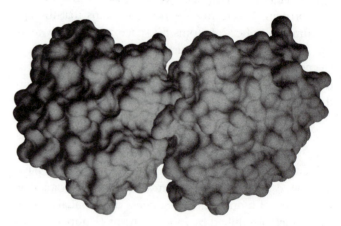

Figure 4-13
The interface between noncovalently bonded polypeptide
chains can be very tight. The packing is as efficient as that of
secondary structural elements in a protein chain. The
example shown here is the hemoglobin $\alpha_1\beta_2$ interface.

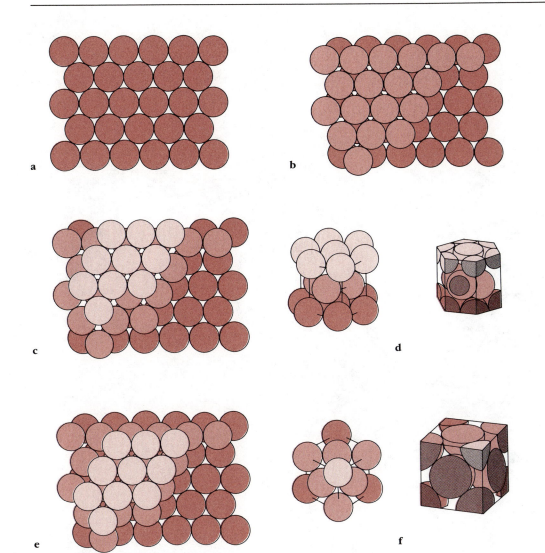

Figure 4-14
a, The first layer of a close-packed array of spheres. **b**, The second layer of a close-packed array. **c** and **e** are the two possible arrangements of a third layer. **c** and **d** show hexagonal close-packing; **e** and **f** show cubic close-packing. They fill space with approximately equal efficiency. Both types of close-packing appear similar, but there are differences. Note that in **c** there are air spaces between the spheres, but in **e** there are none. Cubic close-packing as depicted in **e** and **f** is the most efficient close-packing (74% of space is filled). The interiors of proteins and the stacks of bases in nucleic acids are packed this tightly. (From *Physical Chemistry* [2nd ed.] by P.W. Atkins. ©1982 W.H. Freeman & Co.)

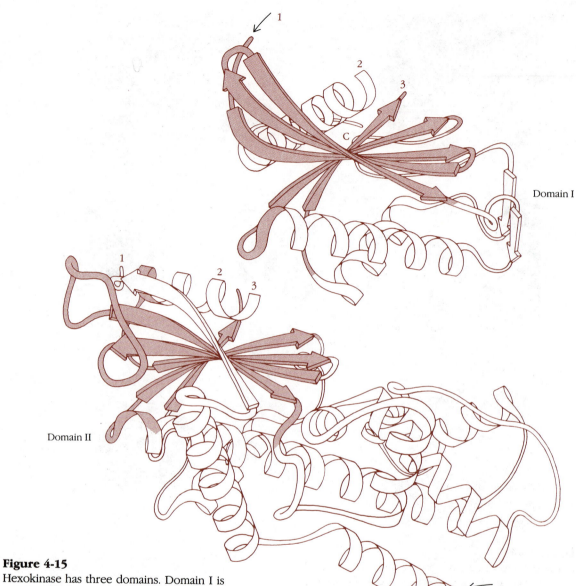

Figure 4-15
Hexokinase has three domains. Domain I is
a β sheet (colored) coated with α helices. A
similar domain (II) with the β sheet shaded
is shown below. Domain I is related to
domain II by a 180° rotation (about a
horizontal axis), which brings the pairs of
points marked *1, 2,* and *3* into
superposition. Domain III is all α helix and
is composed of 116 amino acids. Notice that
domains II and III are fused; that is, they
are a continuous piece of polypeptide
backbone. By carefully comparing the
pattern of β strands in the β sheets, you will
note that the two sheets are similar and may
have arisen from a gene duplication.
(Courtesy Jane Richardson.)

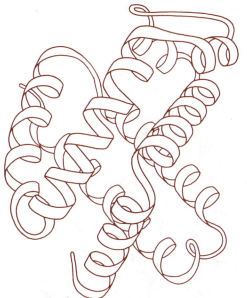

Figure 4-16

The hemoglobin β subunit is constructed from a single domain of eight α helices. These form the binding site for the heme group. (Courtesy Jane Richardson.)

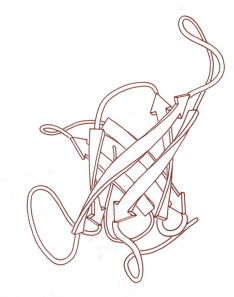

Figure 4-17

A domain of chymotrypsin is constructed of antiparallel β strands to form a barrel-like structure. This folding pattern is common. (Courtesy Jane Richardson.)

Figure 4-18

A domain of lactate dehydrogenase is composed of mixed α and parallel β structure. This domain binds the dinucleotide NADH. (Courtesy Jane Richardson.)

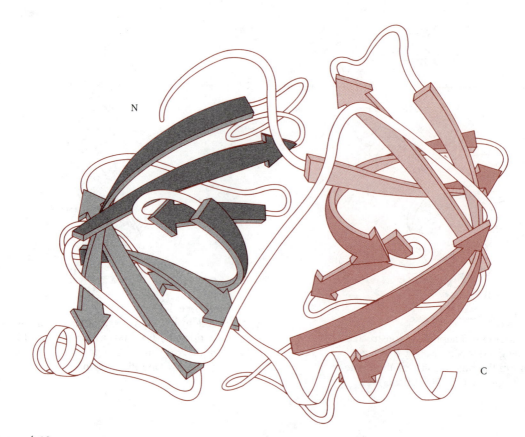

Figure 4-19
Schematic drawing of chymotrypsin showing two similar domains. The β-barrel exteriors are shaded differently in the two domains. Find the homology with Figure 4-17. (Courtesy Jane Richardson.)

Secondary Structure

The types of secondary structure discussed in this book include the α helix, the closely related 3_{10} helix, the antiparallel β sheet, the parallel β sheet, turns, and the collagen helix. More than 90% of all protein molecules are made by simply linking these bits of secondary structure into domains and then subunits to form the entire protein molecule. The individual α helices and β strands that interact to form a domain may be derived from contiguous segments of the polypeptide chain or may be widely separated portions of the polypeptide chain that end up associated as a result of the protein folding into its final tertiary structure. Frequently, proteins appear to be made up of multiple copies of a common domain. Chymotrypsin is an excellent example of this (Figure 4-19): it is composed of two similar domains, each of about 120 amino acids folded into a six-stranded barrel of antiparallel β strands. In fact, approximately one third

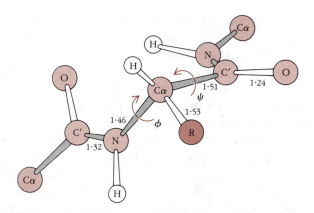

Figure 4-20
The course of the protein backbone is
determined by the two angles θ and ψ for
each amino acid. The bond lengths (in
angstrom units) vary only a little from the
values shown.

of all proteins (of known structure) contain two or more
structurally similar domains. These domains often have
little or no similarity in amino acid sequence but do have
a great deal of structural homology. Figure 4-12 depicts a
portion of an immunoglobulin molecule that is composed of 12 domains of strikingly similar structure.

φ and *ψ* Angles Specify
Protein Structure

X-ray crystallographic analysis of amino acids and groups
of amino acids has shown that the peptide group is a
rigid unit that is nearly always planar due to its partial
double-bond character. Figure 4-20 shows the dimensions that have been worked out from a number of such
studies. Despite the inflexibility of the peptide bond, the
linear chain can still form a large variety of structures.
There is, relative to the peptide bond, free rotation about
the bond between the α-carbon and nitrogen atoms. The
same is true for the α-carbon to carbonyl carbon bond.
The rotational angles about these bonds are designated *φ*
and *ψ* as shown in Figure 4-20. Since there is no rotation
about the peptide bond, because of its partial double-
bond character, *the conformation of the main chain of
the polypeptide is completely determined when the φ and
ψ angles for each amino acid are defined.*

Most *φ*, *ψ* angle pairs are disallowed because of steric
collisions among the side chains and main chain. The *φ*,
ψ angle pair −150°, 150° corresponds to an almost completely extended main chain that has no unfavorable
steric interactions. This is the angle conformation found

in β strands, which self-associate to form the extended β-sheet structure (pp. 76–84). A β sheet is stabilized by main chain hydrogen bonds between pairs of strands.

Other combinations of ϕ, ψ angles allow the NH group of one amino acid to hydrogen bond to the CO in another amino acid within a strand. The formation of these hydrogen bonds causes the polypeptide to fold into a stable, compact structure. The ϕ, ψ pair $-60°$, $30°$ corresponds to a 3_{10} helix, in which a hydrogen bond forms between amino acid (n) and (n + 3) in sequence (pp. 68–76). The ϕ, ψ pair $-65°$, $-40°$ describes an α helix in which hydrogen bonding occurs between every fifth amino acid (pp. 64–67).

The α-Carbon Representation

The polypeptide backbone is formed by only the N, Cα, and C$'$ atoms of each amino acid and is difficult to see when observing an all-atom representation of a protein. The backbone can be followed more simply by just focusing on the α-carbon atoms. Blackwell protein models are α-carbon representations and are convenient for a number of reasons. The α-carbon atoms have a fixed spacing of 3.8 Å along the polypeptide backbone, except in the rare case of *cis* proline. Figure 4-21 shows the *cis* and *trans* configurations of proline. The fixed spacing arises because of a rigid geometry of the peptide link and the fact that rotation is not allowed about the peptide bond. Consequently, the entire peptide link that connects α-carbons is a rigid, well-defined structure of a fixed length.

Since all peptide links are the same, they may be replaced by a straight line or a *virtual bond* for simplicity. An α-carbon representation is one in which the Cα—N, peptide and C$'$—Cα bonds, which compose a zig-zag structure connecting successive α-carbon atoms, are replaced by a straight line. As shown in Figure 4-22, when this is done, the ϕ, ψ angle pairs, which have a physical meaning (corresponding to torsional bond rotations), are replaced by two other angles that have no real physical meaning but that serve to precisely position the

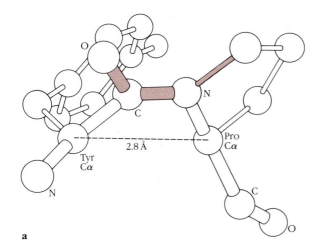

a

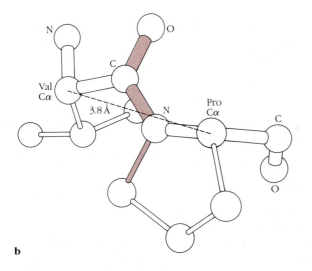

b

Figure 4-21
Proline is the only amino acid that can be (**a**) *cis* conformation (the carbonyl oxygen is on the same side of the peptide bond as the N-to-side chain bond) as well as the usual (**b**) *trans* conformation. The polypeptide backbone is *cis* or *trans* to itself in these cases. The two conformers are shown here for two segments of ribonuclease. Note that the α carbon-α carbon distance is 2.7 Å for *cis* instead of the 3.8 Å for *trans* proline. The *cis* peptide group is highlighted in **a**; the *trans* peptide group is highlighted in **b**.

α-carbon in three-dimensional space. Figure 4-22 shows that these angles are θ, the bend angle between successive virtual bonds, and τ, a torsional angle about the virtual bond.

The angles ϕ and ψ relate atoms within a single residue to one another, whereas θ and τ relate α-carbons to one another. The angle θ is the angle, set at one α-carbon, between the two flanking α-carbons. For example, a θ setting at α-carbon Cα 2 is the angle between α-carbons Cα 1 and Cα 3 (Figure 4-22). The angle τ is the torsional angle set at Cα 2, between Cα 1 and Cα 4

Figure 4-22
The ϕ and ψ angle pairs (**a**) are converted to θ (a bend angle) and τ (a torsional angle) for the α-carbon backbone representation (**b**). A short chain of model parts appears in (**c**).

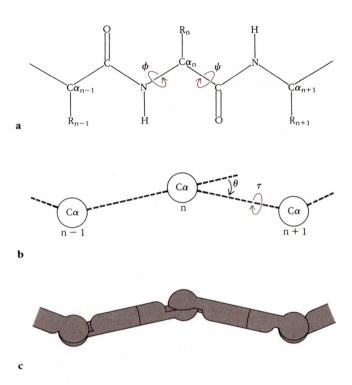

(Figure 10-11). This transformation to θ and τ simplifies the construction of a protein model from a list of Cartesian coordinate values (x, y, z) for each α-carbon atom in the protein to a list of θ, τ for each α-carbon.

Protein structures can be modelled by computers using sets of atomic coordinates generated from x-ray crystallographic studies. If vectors connecting successive α-carbons are drawn, it is a straightforward matter to calculate from atomic coordinates the values of θ and τ that relate any set of α-carbon atoms. It will be seen later that the special secondary structural components of polypeptide chains have well-defined values for the two angles θ and τ, just as for the ϕ, ψ angle pairs. The α-carbon positions in backbone models of proteins are in the exact position they would be in an all-atom representation. The major change is the replacement of the rigid peptide link with a simple straight line. In general, the position of the side chain cannot be determined from the α-carbon positions. However, the regular secondary structures position the side chains in a predictable and regular way. This will be seen in the next chapter.

Summary

The polypeptide backbone is made of a repeating unit of six atoms. These six atoms are called the *peptide link* and include the *peptide bond*, which joins individual amino acids. The peptide link has a defined geometry and is replaced by a straight line in the α-carbon representations.

Individual amino acids are *chiral*, and their assemblies into secondary and tertiary structures also show handedness. Amino acids may be classed simply into two groups: *hydrophobic* and *hydrophilic*. This classification is represented as green and red, respectively, in our modelling system.

Protein folding is determined by physical and chemical forces. Hydrogen bonds and hydrophobic interactions are the major determinants of protein folding. The net result is that soluble proteins have hydrophobic interiors and hydrophilic surfaces. Protein interiors are *close-packed*.

The torsional angles ϕ and ψ define the structure of a polypeptide backbone. Repetitive structures like the α or 3_{10} helix are described by a single pair of ϕ and ψ angles. In the α-carbon representation, the angles θ and τ are used to set the angles that relate one α-carbon to the next.

Suggested Readings

Polypeptide Conformation

IUPAC-IUB Commission on Biochemical Nomenclature (1969): Abbreviations and symbols for the description of the conformation of polypeptide chains. *Biochemistry* 1970; 9: 3471–3479. Presentation of the torsional- and bond-angle definitions that are standardly used in describing polypeptide backbones.

Ramachandran GN, Sasisekharan V: Conformation of polypeptides and proteins. *Adv Protein Chem* 1968; 23: 283–437. Description of the stereochemically allowed conformational space for polypeptide backbones. In it, numerous examples of the so-called Ramachandran plot can be seen. These plots provide the allowed values for the dihedral angles that determine the course of the main chain in polypeptides.

Schultz GE, Schirmer RH: *Principles of Protein Structure.* New York, Springer-Verlag Press, 1979, pp. 18–26. Relatively short discussion of the peptide bond.

Protein Folding

Hol WG, Halie LM, Sander C: Dipoles of the alpha helix and beta sheet: their role in protein folding. *Nature* 1982; 294: 532–536. Discussion of physical properties of secondary structure elements and how they may affect protein structure.

Schultz GE, Schirmer RH: *Principles of Protein Structure.* New York, Springer-Verlag Press, 1979, pp. 10–14: A discussion of side chain properties for the amino acids; pp. 33–36: excellent discussion of hydrogen-bond properties in proteins; pp. 36–41: excellent discussion of the hydrophobic properties of macromolecules and entropic effects on protein folding; pp. 32–33: short discussion of van der Waals contacts as applied to macromolecular structure; pp. 46–52: concise discussion of domains in globular proteins; pp. 42–44: Discussion of the salient aspects of close-packing in macromolecules.

Structural Patterns

Levitt M, Chothia C: Structural patterns in globular proteins. *Nature* 1976; 261: 552–557. Presentation of the definitions of the four common classes of protein tertiary structures and simple schematic representations of each of them. It is a useful reference to classification schemes in protein tertiary structures.

Rose GE: Hierarchic organization of domains in globular proteins. *J Mol Biology* 1979; 134: 447–470. Carefully written article describing attempts to subdivide the tertiary structure of a globular protein into domains. It is worth reading to verify how complicated this issue is.

Sternberg MJE, Thornton JM: On the conformation of proteins: the handedness of the connections between parallel β strands. *J Mol Biology* 1977; 110: 269–283. Description of the handedness that arrangements of secondary structural elements display in three-dimensional space.

Membrane Proteins

Engleman D: Bacteriorhodopsin is an inside out protein. *Proc Nat Acad Sci USA* 1980; 77: 5894–5898. Presentation of proof that a membrane protein is inside out.

Water and Proteins

Hagler AT, Moult J: Computer simulation of the solvent structure around biological macromolecules. *Nature* 1978; 272: 222–226. Sound theoretical analysis of protein-water interactions.

Hildebrand JH: Is there a hydrophobic effect?. *Proc Nat Acad Sci USA* 1979; 76: 194. Readable commentary on semantics and physical chemistry by a then 96-year-old famous chemist.

Rupley J, Yang PH, Tulin G: Thermodynamic and related studies of water interacting with proteins, in *Water in Polymers*, Rowland SP (ed): ACS Symposium series No. 127. American Chemistry Society, 1980, pp. 112–132. Good experimental and theoretical analysis of protein-water interaction.

Side Chain Modifications

Kornfeld R, Kornfeld S: Structure of glycoproteins and their oligosaccharide units, in Lennarz WJ (ed): *The Biochemistry of Glycoproteins and Proteoglycans*. New York, Plenum Press, 1980, pp. 1–34. Thorough discussion of glycoprotein structure.

Krebs EG, Beavo JA: Phosphorylation-dephosphorylation of enzymes. *Ann Rev Biochem* 1979; 48: 923–960. Review of phosphorylation of enzymes with emphasis on regulation of activity.

Packing

Cohen F, Sternberg M: Packing of α helices onto β pleated sheets and the anatomy of $\alpha\beta$ proteins. *J Mol Biology* 1980; 143: 95–128. Analysis of packing interactions.

Richards F: Areas volumes, packing and protein structure. *Ann Rev Biophys Bioeng* 1977; 6: 151–176. Geometric discussion of protein surfaces and volumes.

Serine Proteases

Stroud RM: A family of protein-cutting proteins. *Sci Am* 1974; 231: 74–88. Readable account of the serine protease enzymes.

5

Building Models of Secondary Structure

The tertiary structure of a protein is an assemblage of individual elements of secondary structure. In this chapter you will build and study models of an α helix, 3_{10} helix, parallel β sheet, and three types of turns. It is important to understand these structures fully before embarking on the study of an entire protein. You will also build a model of a collagen molecule, since this abundant protein is composed of a single secondary structure.

The α Helix

Figure 5-1 shows three representations of an α helix, the most common and stable secondary structure found in proteins. The name α *helix* is derived from the classification of x-ray diffraction patterns from natural fibers into the types of α and β. Examples of these are α keratin and β silk fibroin. The all-atom representation in Figure 5-1a shows the position of the side chains as they emanate from the α-carbon atoms as well as the peptide links and associated hydrogen bonds. Figure 5-1b shows just the atoms that compose the backbone of the α helix. Finally, in Figure 5-1c the five bonds of the peptide link have been replaced with the simple virtual bond: this is

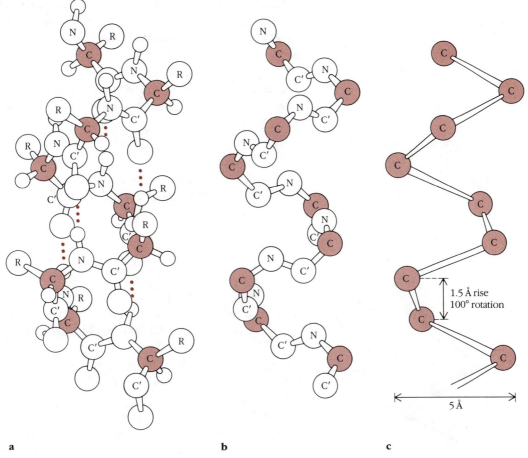

1.5 Å rise
100° rotation

5 Å

a

b

c

Figure 5-1

Three representations of an α helix. The chain runs from the top of the page down in the N to C direction. The hydrogen bonds are denoted by dotted lines in **a**. The side chains are removed to reveal the peptide backbone in **b**. The α-carbons of the backbone are shown in **c**. In all cases, α-carbons are colored. Note that the α-carbons are in exactly the same positions in all three panels.

the α-carbon representation. The peptide groups of amino acids n and n + 4 are linked by hydrogen bonds. The α helix is a very stable cylindrical structure because hydrogen bonds cause the polypeptide backbone to fold into a close-packed structure. The α helix structure is an example of efficient atomic packing. There is no way to coil a polypeptide chain more tightly.

As mentioned earlier, an α helix may be described by a reiterated geometric operation relating successive amino acids. The rotation of 100° *about* and the translation of 1.5 Å *along* the helix axis that relates each residue to the next corresponds to 3.6 (360°/100°) amino acid residues per helical turn. An α-helical repeat unit is 18 amino acids, so that amino acids 1 and 19 are exactly aligned along the helix. The helical repeat unit is much larger than a single helical turn. Five turns of the helix are necessary for a full repeat unit (five turns of 3.6 residues per turn yields 18 residues). The length of the repeat unit is 27 Å, or 18 residues × 1.5 Å translation per residue. The ϕ and ψ angles for an α helix are approximately −65° and −40°. These translate to θ and τ angles of 90° and 230°.

Building the α-Helix Model

The purpose of this exercise is to build a regularly repeating structure, a helix, that shows rotation and translation repeated in three dimensions. After you have built the model, you should note that every fifth residue is approximately aligned along the helix axis but residues 1 and 19 are exactly aligned. You should be able to recognize the right- (or left-) handed nature of the helix. This model is built by constructing 19 amino acid units and setting each of the bend angles (θ) to 90° and each of the torsional angles (τ) to 230°. (See Chapter 3 for instructions.) These angles represent a perfectly regular α helix. You do not need to glue this model. (To build a left-handed α helix, set the θ angles to 90° and the τ angles to 130°).

The completed model appears in Figure 5-2. Most α helices found in proteins contain some irregularity. Sometimes the helices are smoothly curved from end to end like a banana (Figure 5-3) and occasionally two helices are linked by turns so that they take on a kinked appearance. These kinks in protein structures are usually caused by tertiary interactions along the α helix with neighboring amino acids in the rest of the molecule or often by prolines. Since the amino nitrogen of proline is part of a five-membered ring, it has no hydrogen to participate in hydrogen bonding. The polypeptide backbone of the helix is forced to bend at proline residues. The bend changes the distance between the hydrogen–bond acceptor (on residue n) and donor (on residues n + 4) from 2.8 Å, the average length of a hydrogen

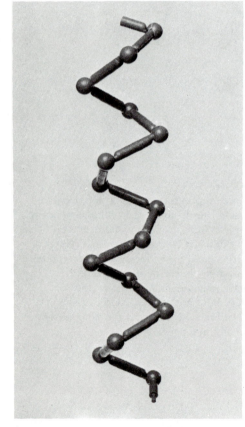

Figure 5-2
Model of α-helix.

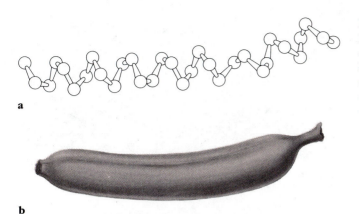

a

b

Figure 5-3
Helices in proteins may be kinked or
smoothly curved as shown here. **a**, This α
helix from the enzyme hexokinase is
smoothly curved like the banana in **b**. This
curvature is due to interactions with the rest
of the protein.

bond, to 3.8 Å (van der Waals distance) when residue n + 4
is a proline. As mentioned before, hydrogen bonds have a
fixed length and geometry, and proline does not allow a
bond with these characteristics to be formed. Proline
thus tends to disrupt regular helical secondary structure
in proteins, although proline may be found at the first
position of a helix. There is a rare example of an α helix
containing proline in the coat protein of tobacco mosaic
virus.

A 19 unit, α-helix backbone is about 5 Å in diameter
and 27 Å long. The α helix shown in Figure 5-4 is a
space-filling representation in which the side chains of
the α-carbons are drawn. The complete α helix is about
10 Å in diameter. Notice that the helix shown in Figure
5-4 has a polar and a nonpolar face (see also p. 134). This
is common, since in proteins one side of the helix usually
faces the solvent water while at least a portion of the
other face packs against and forms part of the hydrophobic core of the protein. Only right-handed α helices
appear in nature. This is a direct consequence of the
handedness of all L-amino acids. If you were to try to
make a left-handed α helix from L-amino acids, you
would find that the side chains would collide with other
atoms and would destabilize the helix. Similarly, D-amino
acids prefer to form left-handed α helices. You can see
what a left-handed α helix looks like by holding your
model or a drawing of an α helix in front of a mirror. The
image you see is one of a left-handed α helix (composed
of D-amino acids). You may also construct a left-handed α
helix according to the instructions on p. 66.

Figure 5-4
Space-filling drawings of an α helix from
flavodoxin. The hydrophilic amino acids are
colored. The side chain distribution gives
one face (**a**) a hydrophilic character and the
opposite face (**b**) a hydrophobic character.

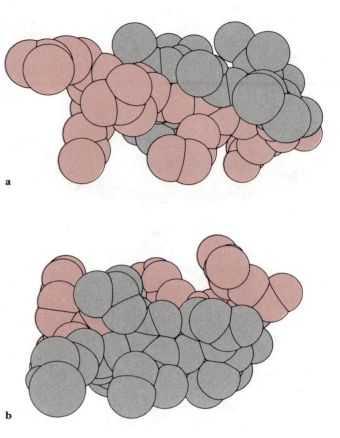

a

b

The 3₁₀ Helix

A common occurrence at the ends of helical portions of
proteins is one or two turns of 3_{10} helix in which every n
and n + 3 amino acids are linked by hydrogen bonds.
The name arises because there are ten atoms in the ring
closed by the hydrogen bond. The 3_{10} helix is only about
5% as abundant as the α helix. No one is certain why
some turns of the helix are 3_{10} and some are α helix.
Since the 3_{10} helix is more extended than the α helix and
slightly less energetically favorable, when a given length
of protein chain is required to stretch further through
space, it might do so by simply switching into the 3_{10}
structure. A 3_{10} helix is shown in Figure 5-5. Covalently
linked amino acids are separated by 2.0 Å along the helix

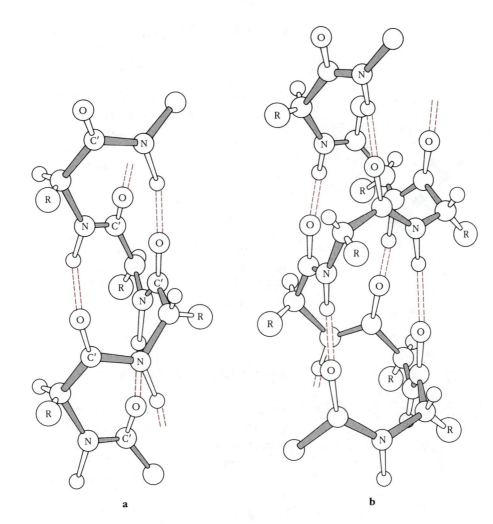

a

b

Figure 5-5
The all-atom representation of a 3_{10} helix
(**a**) is compared with that of an α helix (**b**).
Hydrogen bonds are represented by dashed
lines.

axis. There are only 3 residues per turn in the 3_{10} helix so
the rotation between successive residues is 360°/3 =
120°. Residues 1 and 4 are directly above each other so
the helix repeats after 6 Å (3 × 2.0 Å/residues = 6 Å).
Notice that in the all-atom representation, the backbone
forms a more strained hydrogen bonding network (that
is, the hydrogen bonds are less linear) than in the α
helix. This accounts for the lower stability. The 3_{10} helix
is still right-handed and is described by ϕ, ψ angles of
−60° and −30°, respectively. These translate into θ = 97°
and τ = 262°.

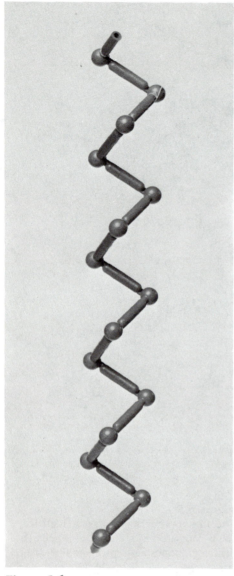

Figure 5-6
Model of 3_{10} helix.

Building the 3_{10}-Helix Model

This construction demonstrates the result of changing the rotational and translational parameters on the form of a helix. Note that the model is thinner and more extended than the α helix model and that every fourth residue is aligned along the helix axis. A 3_{10} helix is built by setting successive τ to 262° and θ to 97° angles for 15 amino acid residues. (To build a left-handed 3_{10} helix, set the θ angles to 97° and the τ angles to 98°.) The 3_{10}-helix model does not need to be glued. You may reuse the parts for other models.

The 3_{10}-helix model appears in Figure 5-6. Notice that this 15-unit helix is also right-handed but is now 28 Å long instead of 21 Å. In cross section (Figure 5-7) the 3_{10} helix is slightly thinner (about 8 Å in diameter with side chains) than the α helix.

The Collagen Helix

The abundant and specialized protein collagen is made from the repeating triplet (Gly-X-Y), where X and Y are frequently proline and hydroxyproline (the hydroxyl group is on the β- or γ-carbon atom of the side chain) as illustrated in Figure 5-8. These particular amino acids preclude the formation of an α or 3_{10} helix, since no hydrogen bonds can form along the chain to stabilize the structure. Instead the repeating triplet forms an extended left-handed helix (Figure 5-9). Hydrogen bonds do not form within such a helix; however, when three individual left-handed helices associate to form a triple-stranded helix as seen in Figure 5-10, hydrogen bonds form between the backbone hydrogen of glycine residues and backbone carbonyl oxygens of proline residues on different strands. A space-filling drawing of collagen is shown in Figure 5-11. The individual helices will not pack together if they are straight, so they twist around each other like the three strands in rope. Three strands of ten triplets each are in the 86 Å long collagen repeat unit. Van der Waals forces and hydrogen bonds hold the

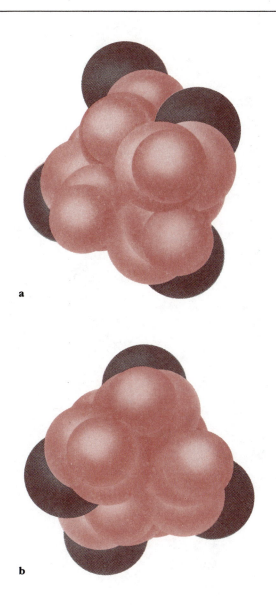

a

b

Figure 5-7
a, Projection down the α-helix axis. The space-filling drawing shows a solid core. **b**, Projection down the 3_{10}-helix axis. Side chain methyl groups are darker.

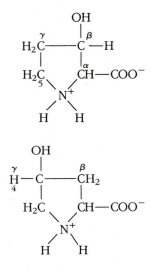

Figure 5-8
The amino acid proline may have a hydroxyl group attached in either the β or γ position. The hydroxyl groups form hydrogen bonds in the collagen molecule.

strands tightly together. However, the strength of the molecule, which makes it suitable for forming tendons and skin, comes from covalent cross-links between side chains on different collagen molecules. Figure 5-12 is an electron micrograph of collagen molecules. The collagen in skin is further stabilized by special covalent links between special amino acid side chains on adjacent triple helices.

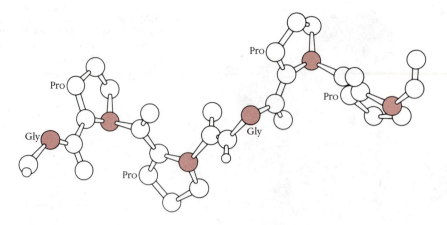

Figure 5-9
A single strand of collagen is a left-handed
helix.

Figure 5-10
Skeletal representation of the triple-stranded
collagen helix. The sequence shown here is
Gly-Pro-Pro. (From Stryer L: *Biochemistry*.
San Francisco, W.H. Freeman & Co., 1981,
p. 189.)

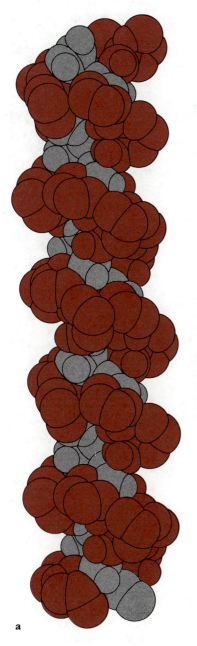

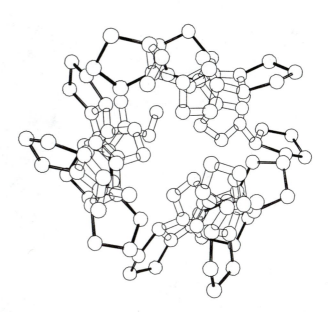

Figure 5-11

a, Computer-generated drawing of collagen shows a close-packed structure with deep grooves. The colored ridges are formed by the proline side chains, found on the outside of the triple helix. Glycine residues are in gray. **b**, Cross section through the collagen helix. The three helix backbones are surrounded by the proline side chains. The helix has an average diameter of about 15 Å. The diameter of the backbone averages 10 Å.

a

Figure 5-12

Top, Electron micrograph of type 1 collagen molecules from skin. The molecules are 15 Å in diameter and 3000 Å long. (×75,000.) *Bottom*, A single triple helical molecule. (×375,000.) (Courtesy Joseph Madri.)

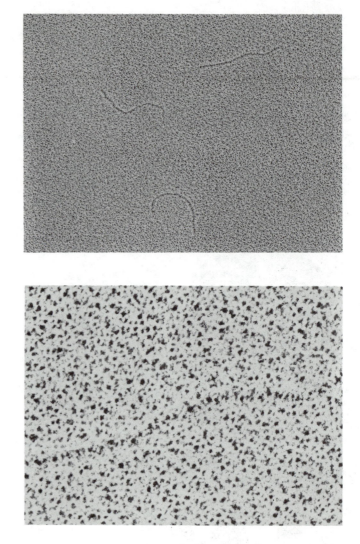

The torsional angles of the helix are $\phi = -51°$, $-76°$, and $-45°$ and $\psi = 153°$, $127°$, and $148°$ for Gly-Pro-Pro (see Table 5-1 for corresponding θ and τ angles). Note that the angles are approximately equal for all three amino acids in the repeat unit. This is because the strands wind around each other and consequently must have similar repeating structures. Because the repeating unit is the triplet (Gly-X-Y), only every third ϕ, ψ pair is really identical.

Table 5-1 Angle settings for collagen helix model

Chain	No.	Color	Amino Acid	$\theta(°)$	$\tau(°)$
1	1	Green	Pro P	56	95
	2	Red	Gly G	59	74
	3	Green	Pro P	71	89
2	1	Green	Pro P	71	89
	2	Green	Pro P	56	95
	3	Red	Gly G	59	74
3	1	Red	Gly G	59	74
	2	Green	Pro P	71	89
	3	Green	Pro P	56	95

Building the Collagen-Helix Model

The purpose of this exercise is to demonstrate the packing of helices in collagen. The helices making up the assembly are curved and show the opposite handedness from that of the assembly. The axes of the individual helices are not straight, as in the α or 3_{10} helix but are, in fact, helical. The curvature of this helix axis is the same as the final assembly. It is always true that in assemblies of helices, the handedness of the individual helices is the *opposite* of that of the assembly. Why?

The collagen helix model is built in two colors using the θ and τ angles in Table 5-1. Three chains should be made, each consisting of 12 amino acids. Make each chain six amino acids at a time and then glue the two segments together. Cut eleven 3.9 cm support rods. The hydrogen bonds are modelled by placing the 3.9 cm supporting rods between Gly 2 of chain 1 and Pro 1 of chain 2, between Gly 3 of chain 2 and Pro 2 of chain 3, and between Gly 4 of chain 3 and Pro 3 of chain 1. Repeat this pattern to add 11 total supports to complete the triple helix of 36 amino acids.

Notice that the individual helices are left-handed and that they are curved and not straight as is the α helix. Figure 5-13 shows the three-stranded collagen helix. The connecting rods represent an approximation of the hydrogen bonds between strands that contribute to the stability of collagen. These are *not* hydrogen bonds be-

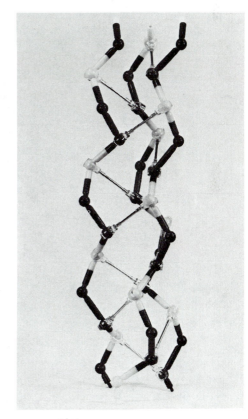

Figure 5-13
Model of collagen.

tween the α-carbons; however the peptide groups of the amino acids are joined by hydrogen bonds in the actual collagen molecule.

These helices are much more extended than α helices (p. 67). The translation distance 2.87 Å relates repeating units; that is, the first triplet in chain 3 is 2.87 Å below the similar unit in chain 1. A 108° rotation will superimpose the first triplet of chain 3 onto the first triplet of chain 1 after the 2.87 Å translation. These parameters may be verified by measurement of the model.

In collagen, a chain of ten amino acids is 28.7 Å long. An α helix ten residues long is 15 Å. The extended and coiled coil collagen structure allows close-packing of the individual chain and one interstrand hydrogen bond per triplet using the N—H of the glycines. The interaction of three otherwise flexible helices grants the final structure significant stability.

Extended Protein Chain: The β Sheet

β sheets in proteins are formed from groups of two to ten or more β strands. Each strand has an average of seven amino acids and may be thought of as very extended simple helices. The β strands in sheets may be parallel, antiparallel, or mixed. A particular protein subunit or domain usually contains either all antiparallel or mostly parallel β strands. Parallel β sheets are often coated by α helices or irregular β strands. Parallel β strands sometimes curve to form a cylindrical barrel.

Along a single β strand, the side chains are positioned alternately up and down. When β strands are assembled into a sheet, the side chains are aligned in rows. All the side chains in a row point up or down. This is true for both parallel and antiparallel β sheets. Figure 5-14 is a schematic representation of a parallel β sheet, in which the side chains are positioned so they point up and down in this manner.

The backbone of the amino acids in a β strand is not completely extended (that is, $\psi = \phi = 180°$), since this is

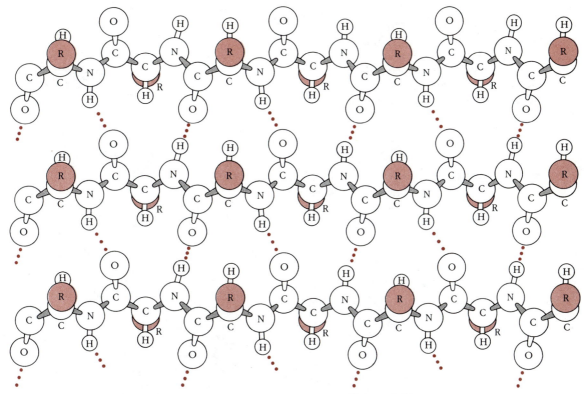

Figure 5-14
The side chains (colored) in this idealized parallel β sheet are pointing alternately above and below the plane of the page.

not energetically favorable. A typical stable β strand has torsional angles of $\phi = -120°$, $\psi = 120°$. These correspond to $\theta = 55°$ and $\tau = 10°$. Figure 5-15 shows three strands in a perfectly regular antiparallel β sheet. However, in most naturally occurring structures, slight steric conflicts between the side chains cause a systematic variation in the α-carbon positions as the strand proceeds. Each amino acid of a single strand is slightly rotated to the right (or clockwise), as you look end on from the N to C direction. This rotation in most proteins is small and variable. Another feature that determines β-sheet structure is that the distance between neighboring α-carbons on adjacent strands is almost always 4.7 Å. The consequence of the systematic twisting of the single strand to the right and the fixed spacing of adjacent β strands is that the strands assemble into a β sheet that is also twisted. β sheets have a handed character. Strand 2 is rotated 10°, like a propeller, relative to strand 1. The side

Figure 5-15
Three-stranded antiparallel β sheet from concanavalin A. **a**, α-Carbon representation. **b**, All-atom representation of the sheet depicted in **a**. Note the alternating up-and-down positions of the R groups above and below the plane of the paper and the pattern of hydrogen bonds.

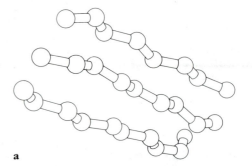

a

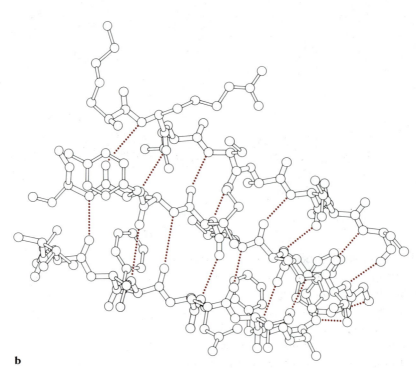

b

chains would collide if there were no rotation. Because the amino acids are all L-isomers, the strands always rotate in the same direction. Strand 3 is rotated 10° from strand 2 and 20° from strand 1. Figure 5-16 presents a view of a β sheet in flavodoxin. There are a few examples of perfectly flat planar β sheets in proteins (for example, in alkaline phosphatase and glutathione reductase).

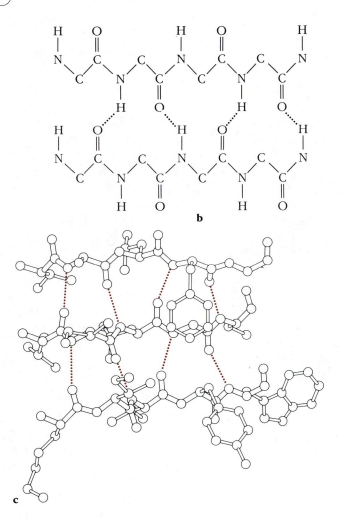

a

Figure 5-16

β sheets in proteins are usually twisted.
a, The twist is obvious in this α-carbon
representation of a parallel β sheet from
flavodoxin. **b**, Schematic version of a
parallel β sheet. **c**, All-atom representation
of the same region of flavodoxin.

b

c

Building a Parallel β-Sheet Backbone Model

The purpose of this exercise is to demonstrate the twist of β strands and a β sheet. You will see that the β-sheet twist can be described in two ways. The twist arises as a result of a slight right-handed helical twist of individual β strands. This twist has profound implications on the packing of β sheets and α helices in the tertiary structure of proteins.

To build a β sheet, assemble 24 amino acid units into four strands of six each. Set the θ angles to 55° and τ angles to 10° for each strand.

Each strand has a right-handed twist, which is evident when you look down the strand from the N terminus. The twist you see is approximately 30°. This is measured as the torsional angle (τ) relating the two "bonds" between α-carbons 0 and 1 and 6 and 7. Using this rough approximation, we can calculate that a β-strand helix would repeat about every 72 amino acids ($360°/30° \times 6$ amino acids = 72). The rise per residue is approximately 3.3 Å in this case.

After the four individual strands are assembled, add nine connecting links (2.7 cm), three between each of two parallel strands, as shown in Figure 5-17. You can shape the structure into its naturally twisted form by laying the fully assembled unit on a table and making certain that the dips and rises in the α-carbon strands are approximately perpendicular to the surface. Pick up the structure, and holding strand 1 (the one farthest from you) in one hand, twist strand 2 clockwise as far as it will go (about 10°). Holding strand 2, rotate strand 3 similarly clockwise about 10°. Finally, holding strand 3, rotate strand 4 clockwise about 10°. Adjust the supports and strands carefully so that the sheet is twisted as is shown in Figure 5-18. (The twist can be exaggerated in the model by removing the two supports at the extreme N terminus of the sheet and the one support at the C terminus as in Figure 5-17). If you wish, you may secure the structure now by gluing the connectors and support rods. These supports represent the actual distance between α-car-

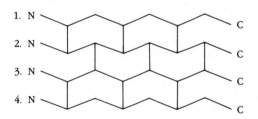

1. N
2. N
3. N
4. N

C
C
C
C

Figure 5-17
Pattern of connecting links in the β-sheet model. The numbers correspond to the strands described in the building instructions.

bons and their respective side chains on adjacent strands. The supports do *not* represent the hydrogen bonds between backbone atoms.

Note that the structure in Figure 5-18 has a particular handedness. If you look down the strands, you will see that the structure (the assembly of strands) shows a right-handed twist that is observed whether you look down the strands from either the N or C terminus. However, if you view the model perpendicular to the strands, strand 2 shows a left-handed twist with respect to strand 1, and so on for the other two strands. Thus the completed β sheet shows a left-handed twist when viewed perpendicular to the strands, and a right-handed twist when viewed along the strands. All β sheets in proteins have this same approximate twist. The six or seven strands in a parallel β sheet will typically twist through 90°.

Interactions of α Helices and β Sheets

A major consequence of the β sheet twist is that α helices may associate with twisted sheets in a limited number of ways. Try to lay the backbone model of the α helix that was constructed in association with the twisted sheet. You will see that when it is placed with its long axis along the diagonal of the β sheet, it does not form a uniform close contact along the sheet. It physically collides with the center or edge, depending on whether it was placed at positive 45° or negative 45° with respect to the strands. In most proteins where α helices and β sheets associate to form a close-packed structure (Figure 5-19), the helices are closely aligned (between 0° and 10°) along the axis of the strands. Figure 5-20 shows three possible ways to associate an α helix with a twisted β structure.

Antiparallel β Sheets

The individual strands in antiparallel β sheets show more irregular associations than parallel β sheets. As shown in Figure 5-21, they are sometimes nearly straight, whereas in Figure 4-17, the strands twist so wildly that six of

a

b

Figure 5-18
β-sheet model. **a,** Left-handed twist of the β sheet. **b,** Right-handed twist is obvious in this end-on view.

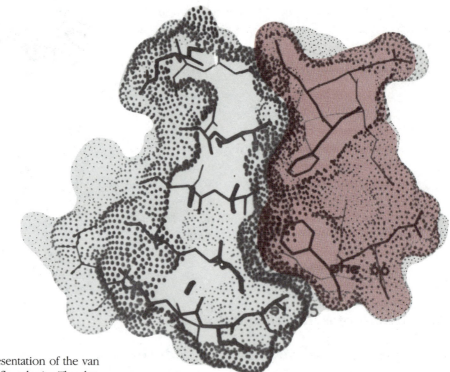

Figure 5-19

Solvent accessible representation of the van der Waals surface from flavodoxin. The dots represent the closest a water molecule could come to the surface of the protein. Two α helices (*right*, colored) are shown packing against a sheet (*left*). Note that this interface is closely packed.

a

b

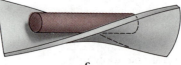

c

Figure 5-20

a, The helix-sheet interaction is best when the helix is aligned along the sheet in the same direction as the β strands. **b** and **c**, The straight α helix does not pack tightly against the curved sheet.

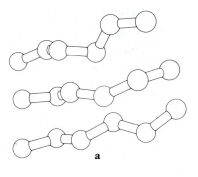

Figure 5-21

Antiparallel β sheet from superoxide dismutase. (**a**) α-Carbon representation. (**b**) Schematic version of an antiparallel β sheet. (**c**) All-atom version of the sheet depicted in **a**. Compare the hydrogen bond pattern with that in Figure 5-16.

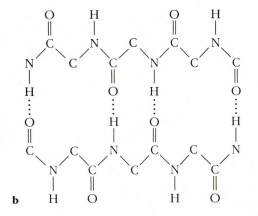

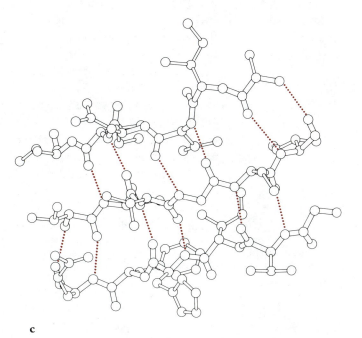

them form a closed β-barrel structure (these structures infrequently form from parallel β strands). The β-barrel structure is a common domain in the large family of serine proteases. It must be a very stable structure since this class of enzymes is relatively stable and suited to their extracellular environments.

Turns: Changes of Chain Direction in Proteins

The α-helix and β-strand secondary structural elements are linked by turns. Although turn structures are highly variable, they are considered regular secondary structural elements. Turns are most often located at the surface of proteins and are usually composed of hydrophilic residues and frequently glycine and proline. About one in five residues in a typical small protein is involved in a turn. Figure 5-22 shows two reverse turns in a βαβ unit. A wide variety of φ and ψ angles are observed in turns. Two-thirds of all turns seen in proteins fall into three well-defined conformational classes. These are shown in Figures 5-23, 5-24 and 5-25. Type I has the approximate conformation of a turn of 3_{10} helix though the φ and ψ angles are distorted from the values for a 3_{10} helix. Type II is similar to type I, but the peptide group is flipped (the carbonyl oxygen points in opposite directions). In this case the oxygen atom of residue 2 points at the side chain of residue 3. Thus glycine is the only amino acid that is found in position 3 of this type of turn. Type III is a perfect piece of a 3_{10} helix formed by four consecutive amino acids. Thus type III is really very much like a type I turn. The principal difference is that the τ angle at residue 2 of the turn is slightly larger (20°) in type III. The enzyme carboxypeptidase contains examples of all three kinds of turns.

Turns are rarely found in the interior of proteins. The reasons for this have already been mentioned. For a peptide backbone to turn, there must be some peptide groups whose NH and CO groups cannot form hydrogen bonds as they do in the α helix and β sheet. The close-packed nature of a protein's interior generally excludes

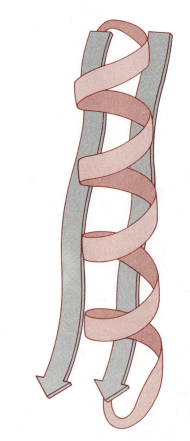

Figure 5-22
Unit of βαβ showing two reverse turns. The spatial arrangement of these secondary structure elements is right-handed. This βαβ structural unit is commonly found in proteins.

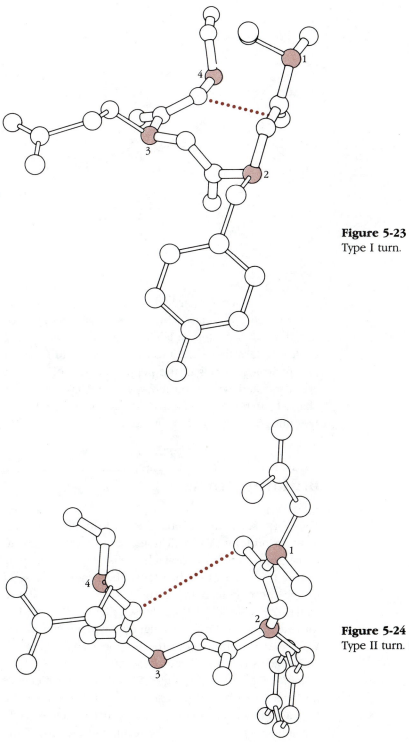

Figure 5-23
Type I turn.

Figure 5-24
Type II turn. Glycine is in position 3.

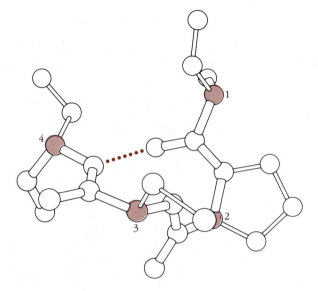

Figure 5-25
Type III turn is a portion of a 3_{10} helix.

water, and since unpaired hydrogen-bonding partners are energetically unfavorable, suitable partners must be found in other protein side chains or buried water molecules. In fact, there are a few buried turns in proteins. The hydrogen bonds of the peptide groups of the turn are associated with a buried water molecule and side chains from the protein. Examples are found in lysozyme (residues 54 to 57, p. 174) and trypsin.

Building Turns

The purpose of this exercise is to learn to recognize turns in proteins. You will note that turns can serve to relate secondary structural elements at widely varying angles, but in all cases they cause the general course of the polypeptide to bend. Turns can be seen in a Blackwell protein model fairly easily (see Chapters 8, 9, and 10 for examples) and can be assigned a type equally easily. You need simply to note the torsional angle between amino acids two and three of the turn.

Tables 5-2 to 5-4 list angles for different segments of the protein carboxypeptidase. Angles are given for the four amino acids that compose the turn as well as flanking amino acids at the N and C termini of the turn. Table

Table 5-2   Angle settings for type I-turn model

No.	Color	Amino Acid	$\theta(°)$	$\tau(°)$
36	Green	Leu L	44	53
37	Green	Gln Q	67	358
38	Green	Ile I	91	123
39	Green	Gly G	32	99
40	Green	Arg R	54	64
41	Red	Ser S	51	88
42	Red	Tyr Y	97	222
43	Red	Glu E	82	82
44	Red	Gly G	84	328
45	Green	Arg R	69	66
46	Green	Pro P	64	43
47	Green	Ile I	76	24
48	Green	Tyr Y	52	7
49	Green	Val V	55	35

Table 5-3   Angle settings for type II-turn model

No.	Color	Amino Acid	$\theta(°)$	$\tau(°)$
84	Green	Lys K	93	236
85	Green	Lys K	85	231
86	Green	Phe F	89	233
87	Green	Thr T	92	221
88	Green	Glu E	74	234
89	Red	Asn N	84	277
90	Red	Tyr Y	72	192
91	Red	Gly G	83	236
92	Red	Gln Q	83	233
93	Green	Asn N	58	15
94	Green	Pro P	71	233
95	Green	Ser S	95	222
96	Green	Phe F	82	230
97	Green	Thr T	93	236

5-2 lists the angles required for constructing a typical type I turn. These angles are for carboxypeptidase residues 36–49. The turn is formed by residues 41–44 (serine, tyrosine, glutamic acid, and glycine). Figures 5-26, 5-27, and 5-28 show the three turns. The models of turns do not need to be glued.

Table 5-3 lists the angles for constructing a type II turn. These are for carboxypeptidase residues 84-97, with the turn formed at 89-92 by aspargine, tyrosine, glycine, and glutamine. An example of a type III turn (3_{10} helix) is

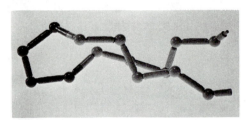

Figure 5-26
Model of a type I turn from
carboxypeptidase.

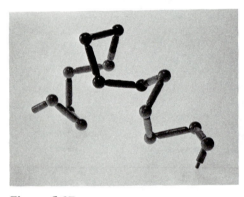

Figure 5-27
Model of a type II turn from
carboxypeptidase.

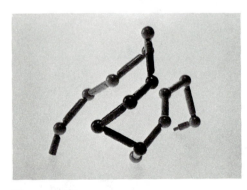

Figure 5-28
Model of a type III turn from
carboxypeptidase.

Table 5-4 Angle settings for type III-turn model

No.	Color	Amino Acid	$\theta(°)$	$\tau(°)$
154	Green	Ala A	79	253
155	Green	Gly G	91	223
156	Green	Ala A	31	60
157	Green	Ser S	48	33
158	Green	Ser S	86	284
159	Red	Ser S	70	30
160	Red	Pro P	95	240
161	Red	Cys C	92	287
162	Red	Ser S	60	37
163	Green	Glu E	96	232
164	Green	Thr T	92	340
165	Green	Tyr Y	69	364
166	Green	His H	77	49
167	Green	Gly G	40	113

seen in residues 159-162 (serine, proline, cysteine, and serine) in carboxypeptidase. Table 5-4 presents the angles for constructing the region of carboxypeptidase from residues 154 through 167. Residue 160 (proline) in this turn can be thought of as initiating the 3_{10} helix.

You should note that the turn structures look different and it is difficult to see any consistent structural feature. The three examples show secondary structural elements intersecting at a variety of angles. The single feature you should note is that the turns linking these pieces of secondary structure serve to *change the general direction* of the polypeptide backbone.

The type of turn is assigned by the angle at which the secondary structural elements diverge. Look at the four α carbons that make up the three turn structures. Look down the torsional bond between α-carbons 2 and 3 in the turn. For type I turns, the angle you see between the two adjacent torsional bonds is approximately 40° (222° − 180° = 42°). For type II turns this angle is approximately 10° (192° − 180° = 12°), and for type III turns the angle is approximately 60° (240° − 180° = 60°). Thus, the four α carbon atoms are approximately coplanar in type II turns. Type I and type III have increasing opening angles (τ for α-carbon 2). Note that the direction of this angle is constant though the amount will vary with the particular turn.

Summary

In an α helix, α-carbons n and n + 4 are linked together via the hydrogen bond between the CO and NH of the peptide groups. In the 3_{10} helix, the α-carbons n and n + 3 are linked. The α helix is much more common and is shorter and thicker than the 3_{10} helix.

Collagen is a right-handed triple helix composed of three left-handed helical strands of proline and glycine residues. The individual helical strands are very extended, and the helix repeats itself every 86 Å compared to every 27 Å for an α helix.

The individual strands in a β sheet are arranged either parallel or antiparallel. Neither kind of β sheet is flat. Most parallel β sheets have a slight right-handed twist. Antiparallel sheets tend to be more variable and frequently form a closed barrellike structure.

Secondary structure elements are linked into tertiary structure by turns that change the direction of the polypeptide backbone by angles between 90° and 180°. Even the largest, most complicated protein structures can be divided into these individual structural components.

Suggested Readings

Helices and Sheets

Schulz G, Schirmer RH: *Principles of Protein Structure*. New York, Springer-Verlag Press, 1979, pp. 67–78. Discussion of secondary structure elements.

Collagen Structure

Okuyama K, Okuyama K, Arnott S, et al: Crystal and molecular structure of a collagen-like polypeptide (Pro-Pro-Gly)$_{10}$. *J Mol Biology* 1981; 152:427–443. Description of a polymorphic structure.

Piez KA: Structure and assembly of the native collagen fibril. *Connect Tissue Res* 1982; 10:25–36. Model for the structure of the collagen fibril.

Piez KA, Trus BL: Sequence regularities and packing of collagen molecules. *J Mol Biology* 1978; 122:419–432. Detailed discussion of molecular packing in collagen.

Turns

Rose G, Young WB, Gierasch LM: Interior turns in globular proteins. *Nature* 1983; 304:654–657. Discussion of buried turns in proteins.

6

Nucleic Acid Structure

Nucleic acids are used to carry genetic information and to translate it into protein. Their structures must be specifically suited to these functions. Base-pairing between complementary bases on the same or different strands is the major structural feature of all nucleic acids. In the DNA structures, the complementarity between bases on different strands allows fastidious maintenance of the sequence of bases. In tRNA, double-stranded regions form from a single strand of nucleic acid. These double-stranded *stems* fold into the characteristic L-shaped tRNA tertiary structure. Hydrophobic base-stacking interactions and charge separation between the backbone phosphate groups stabilize the final secondary or tertiary structure.

The DNA Molecule

DNA is the molecule responsible for the richness of life forms and for passing the assembly instructions from one generation to the next. DNA carries the information of heredity in all cells. The features that make DNA suitable

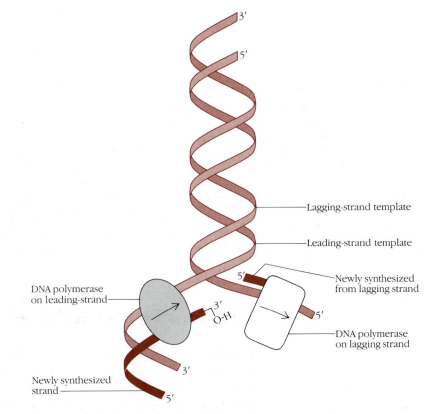

Lagging-strand template

Leading-strand template

Newly synthesized
from lagging strand

DNA polymerase
on leading-strand

DNA polymerase
on lagging strand

Newly synthesized
strand

for this function are its relative chemical and physical stability and *complementarity*. DNA is always double stranded, and each strand contains the necessary information to build the other. The structure of DNA is the basis for this elegant symmetry. Each strand is made from just four bases—adenine (A), guanine (G), cytosine (C), and thymidine (T)—linked to the constant deoxyribose-phosphate backbone. The exclusive noncovalent coupling of A with T (only) and G with C (only) and the twofold axis that relates the two chains of DNA are the physical basis of the complementarity. DNA can, with help of the DNA-replicating enzymes, replicate itself by strand separation. This is shown in Figure 6-1. The actual replication is a very complex process; only one of the two strands can be replicated continuously as the replication apparatus moves along the DNA helix.

Figure 6-1
DNA replication. DNA polymerase synthesizes daughter DNA strands (colored) from the parent strands as the helix is unwound. Synthesis on the leading strand is continuous, while synthesis on the lagging strand is discontinuous, leaving short gaps that must be sealed later. At least seven proteins are involved in the process depicted.

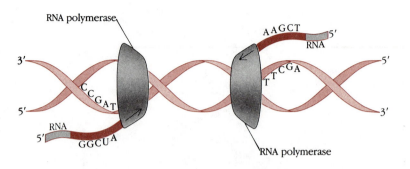

Figure 6-2

Transcription. RNA polymerase (open trapezoids) can copy either DNA strand into mRNA. Newly synthesized RNA is made from the 5′ to the 3′ direction. The DNA is read from 3′ to 5′. Newly made RNA is in color.

The information for building the proteins is copied to messenger RNA for protein synthesis by RNA polymerase (Figure 6-2). In all cells, mRNA (messenger ribonucleic acid) translocates the genetic information from the nucleus (or nuclear region) to the cytoplasm where protein synthesis occurs.

The Transfer RNA Molecule

Messenger RNA carries the information that determines the amino acid sequence of a protein in the form of triplets of nucleotides called *codons*. This information must be translated from the genetic code into an actual protein molecule. Ribosomes contain the enzymatic machinery that links individual amino acids into peptides, but the specific amino acid indicated by the mRNA must be placed into the correct site in the ribosome before linking can occur. The tRNA molecule serves as the adapter between the nucleic acid (mRNA) and an amino acid. Each codon specifies an amino acid, and each tRNA has a site to which a specific amino acid is attached. Each tRNA also contains a site, the *anticodon*, that reads a specific codon. Thus there is a tRNA molecule for each of the 61 codons that specify amino acids (that is, there are no tRNAs for stop codons). In the ribosome, the tRNA with its cognate amino acid attached binds to the appropriate codon on the mRNA. This binding involves the formation of three Watson-Crick base pairs between the codon and the anticodon on tRNA. Figure 6-3 shows the major components of the translation process.

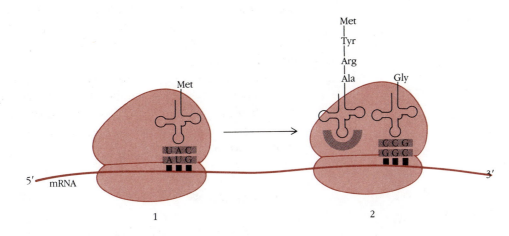

The Nucleic Acid Backbone

Like proteins, nucleic acids can be simply modelled by using virtual bonds to link the individual phosphate groups of the phosphodiester backbone. However, since the phosphate-to-phosphate distance is dependent on the conformation of the backbone, a fixed length for these virtual bonds cannot be set. The consequence of this is that the scale of the nucleic acid backbone model varies with the phosphate–phosphate distance. The scale is different from that of the protein models and is also slightly variable for the nucleic acid models described here. The scale of the nucleic acid models is approximately two-thirds that of the protein models (0.6 cm/Å as compared to 1 cm/Å for the protein models). This means that the nucleic acid models are two-thirds the size they should be relative to the protein models.

The individual nucleotide residues that make up nucleic acid molecules are linked by *phosphodiester bonds*. The phosphodiester bond, seen in Figure 6-4, is the phosphate group that connects the 3′-hydroxyl of one ribose to the 5′-hydroxyl of the next ribose. Unlike the peptide link, the *phosphodiester link* is not rigid and geometrically well-defined. Consequently, six torsional angles, seen in Figure 6-5, define the course of a phosphodiester backbone as compared with the two torsional angles ψ and ϕ that define the structure of a polypeptide backbone (p. 57). The angle χ is the torsional angle of the

Figure 6-3
Translation. The ribosome is the site where codons in mRNA are read by tRNAs and translated into protein. Shown here are two ribosomes on a mRNA. Ribosome 1 has just bound and is reading the first codon, and ribosome 2 has synthesized a short polypeptide.

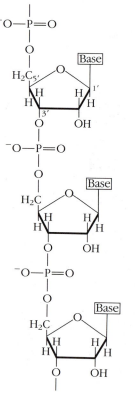

Figure 6-4
The phosphodiester bond links riboses through the 5'- and 3'-carbons. This shows part of an RNA molecule.

bond joining the base and ribose. Although they are all different, certain torsional angles are dependent on each other. In no case, however, are they precisely fixed as in the rigid peptide link.

The torsional and bend angles used to construct the backbone models are not simply related to the angles that define the phosphodiester link. The angles θ and τ are calculated from the atomic positions of the phosphorous atoms.

Nucleotides

Nucleic acids are long, linear polymers of repeating subunits (Figure 6-6). The monomers that make up nucleic acids are called *nucleotides*. Nucleotides are purine or pyrimidine bases attached to a sugar-phosphate. The phosphate is what differentiates nucleotides from nucleosides. The sugar is either ribose, in ribonucleic acids (RNAs) or deoxyribose, in deoxyribonucleic acids (DNAs). Deoxyribose lacks a hydroxyl group that is found on ribose in the 2' position of the sugar. DNA differs from RNA in a second way. The base thymine (T) in DNA is replaced in RNA by uracil (U). Thymine has a methyl group in the 5 position of the pyrimidine ring, while uracil has a hydrogen atom.

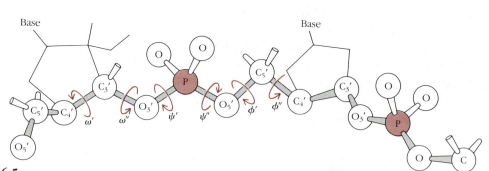

Figure 6-5
Six interdependent torsional angles determine the conformation of a nucleic acid chain.

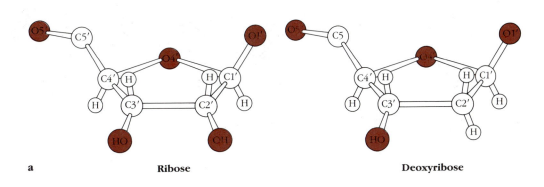

a   **Ribose**   **Deoxyribose**

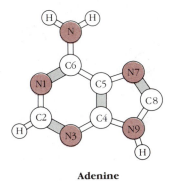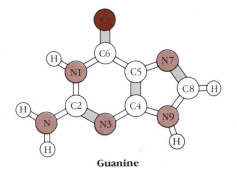

b   **Adenine**   **Guanine**

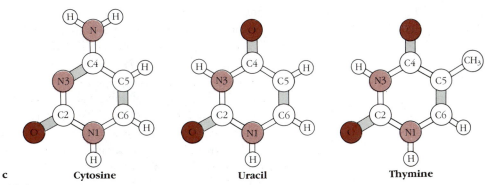

c   **Cytosine**   **Uracil**   **Thymine**

Figure 6-6

Bases and sugars of nucleic acids. Double bonds are shaded. Nitrogens are light-colored; oxygens are darker colored.

Figure 6-7
a, A Watson-Crick A-T base pair has two hydrogen bonds. **b**, The G-C base pair has three hydrogen bonds. Color scheme is the same as in Figure 6-6.

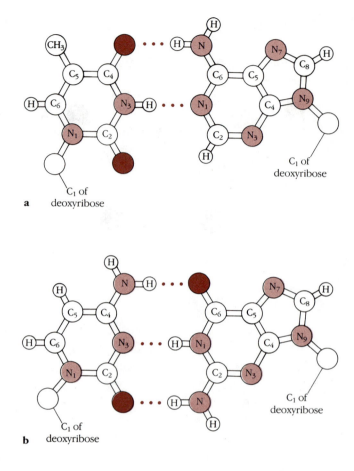

Many properties of nucleic acids are determined by the strong electric charges on the molecule. The phosphate group in the phosphodiester backbone is a strong acid, hence the name *nucleic acid*. Each phosphate in the backbone is negatively charged at pH 7.

Another salient feature of nucleic acids is that the bases have strongly polar nitrogen and oxygen atoms that are able to participate in hydrogen bonding. The most well-known hydrogen-bonded structures in DNA, the Watson-Crick base pairs, are shown in Figure 6-7. The consequence of Watson-Crick base-pairing is that the two strands of the DNA double helix are complementary to each other. The analogous complementarity is found in the double-helical regions of RNA. However, the single

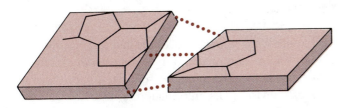

Figure 6-8
Two planar bases related by propeller twist. Note that they are not coplanar and the hydrogen-bonding pattern is twisted slightly.

feature distinguishing RNA from DNA in terms of tertiary structure is that RNA molecules are usually single stranded; thus regular secondary structure in RNA is only seen when a single strand folds back and pairs with itself, forming a double helix. Elsewhere, tertiary interactions (nonbonded interactions that are not secondary structure) between the bases and the riboses are responsible for giving RNA molecules a discrete three-dimensional structure. You will see excellent examples of these interactions in tRNA.

The final structure of double-stranded nucleic acids is greatly stabilized by noncovalent *base-stacking interactions*. These are due to van der Waals attractions and the hydrophobic effect. The bases found in Watson-Crick base pairs do not always lie in a plane. In A- and B-DNA the bases are twisted relative to one another like propeller blades. This type of twist is aptly called *propeller twist* and is depicted in Figure 6-8. The hydrogen bonds between the bases are slightly strained, but hydrophobic base-stacking is maximized.

The forces and principles of the molecular packing that produce base-stacking interactions are analogous to those described for the interior of proteins. The bases are flat, ring structures and are mostly nonpolar. The polar atoms that can participate in hydrogen bonding are those that are involved in Watson-Crick base-pairing and those that are directed into the solvent to hydrogen bond with water molecules or ions. The remainder of the base is hydrophobic, and the bases associate as any hydrophobic molecules. The bases associate in stacks at the center of double-stranded nucleic acids, forming a close-packed interior. Figure 6-9 shows a space-filling diagram of B-DNA with the van der Waals surface as water or ions would "see" it.

Figure 6-9
Computer-generated image of the B conformation of DNA. Note that the charged phosphate groups are on extended ridges and are directed into the solvent. Chemically active groups are shaded (see Figure 2-9). (From Connolly M: *Science* 1983; 221:709–713. Copyright 1983 by the American Association for the Advancement of Science.)

Summary

Nucleic acids are responsible for the faithful transmission of information from generation to generation and from DNA to protein. DNA must be replicated accurately at each cell division. Its information is transcribed into mRNA, which is then translated into a polypeptide, using tRNA as a nucleic acid-protein intermediate. In this sense, nucleic acids are an important structural component of cells as well as being the bearers of genetic information.

Nucleic acids are made of a phosphodiester backbone of sugars linked by phosphate groups. The side chains are purine and pyrimidine bases. An important interaction between side chains is the well-known Watson-Crick base-pairing, which involves hydrogen bonding between two specific bases. The structure of nucleic acids is governed primarily by Watson-Crick base-pairing, although hydrophobic and van der Waals interactions known as base-stacking interactions also play a role in their stability. The structure of the phosphodiester backbone is governed by six angles. In the modelling system in this text, the angles θ and τ are used to specify the structure of the backbone.

Suggested Readings

Alberts B, Bray D, Lewis J, et al: *Molecular Biology of the Cell.* New York, Garland Publishing Co., 1983, pp. 91–110. The finest textbook in print on cell biology.

Dickerson RE: The DNA helix and how it is read. *Sci Am* 1983; 249:94–111. Review of the modern findings in the field of DNA structure that is well illustrated, well written, and simple to understand. Presentation of the important structural details of base roll and propeller twist that have important sequence-dependent variation in helical DNA structure.

Rich A, Kim SH: The three-dimensional structure of transfer RNA. *Sci Am* 1978; 288:52–62. Presentation of the x-ray structure of tRNA.

7

Building Models of Nucleic Acids

There are at least four distinct forms of double-helical DNA: A-DNA, B-DNA, C-DNA, and Z-DNA. The A, B, and Z forms are considered in this chapter. The C conformation, which is similar to B-DNA, is not discussed here. The existence of each of these structures has been confirmed by x-ray crystallographic analysis. In most cases, the structures show overall regularity analogous to that seen for β strands and α-helices in proteins.

The two-fold symmetry axes that relate the two strands of the DNA double helix are of central importance to its structure-function relationship and so to life. Many conformations of DNA are known to exist in isolated DNA or in complexes of DNA with proteins, and all of them show the two-fold symmetry axes. The backbone models clearly display the simple but complex symmetry of DNA. The structure of tRNA shows some double-stranded regularity but also displays tertiary interactions that cause the single strand of RNA to fold into a tightly bent double helix.

B-DNA Conformation

The most commonly depicted form of DNA is the B form, the structure worked out by Watson and Crick in 1953. These investigators showed that DNA is composed of two

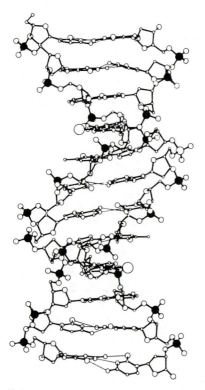

a

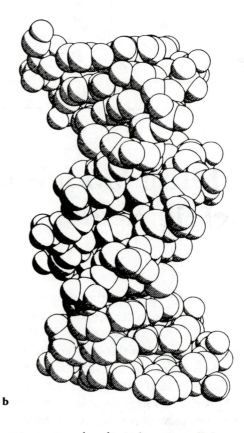

b

Figure 7-1
Skeletal (**a**) and space-filling representations (**b**) of B-DNA. Note the propeller twist of the base pairs in (**a**). (From Dickerson RE, Drew HR, Connor BN, et al: *Science* 1982; 216:475–485. Copyright 1982 American Association for the Advancement of Science.)

complementary strands aligned antiparallel to each other. It is presumed, but not established, that a similar form of DNA is found in nucleosomes, the protein-DNA complexes, which are the primary structural component of all chromatin.

There are ten nucleotide pairs in the B-DNA repeat unit. Figure 7-1 shows two representations of the DNA molecule. Notice that in this molecule the base pairs are perpendicular (within 5°) to the helix axis and that the radius of the helix is roughly 10 Å (twice that of an α helix). The base pairs in the B-DNA helix show about 12° of propeller twist. The base-stacking is greater if the bases have this much twist than if the bases are coplanar. The phosphate groups and sugars that compose the backbone are on the outside of B-DNA, giving it a highly charged surface. In solution, the phosphate groups are presumed to be interacting with water or positively charged ions.

Building the B-DNA Model

The purpose of this exercise is to compare the helical structure of B-DNA with other helices. You should note that B-DNA is a two-start or double-stranded helix composed of antiparallel strands. These features are essential for the complementarity that allows replication and transcription. You will also learn about the symmetries that are generated by two-start, antiparallel helices.

You can build a double helix of B-DNA from two strands of 24 alternating red and green nucleotide units. This model will take about 1 hour to build. Set angles as shown in Table 7-1. Align and intertwine the two chains so that red 1 of chain 1 is opposite green 24 of chain 2. Clip on the twenty-four 8.7 cm rods with connectors and extend or compress the model so that the major groove is 10.6 cm and the minor groove is 7.5 cm. The rods are almost perpendicular to the helix axis and represent Watson-Crick hydrogen bonds. Remember, the two chains are antiparallel. Figure 7-2 shows the completed model of B-DNA.

Note that the DNA backbone model (which has a scale of 0.59 cm/Å) contains ten base pairs per turn. The rise per residue is 3.4 Å, so one helical turn is 34 Å. The molecule has a major and minor groove (Figure 7-2). The distance between phosphates directly above each other in the minor groove is only 12.8 Å (7.5 cm), whereas the distance between phosphate groups directly above each other in the major groove is 18.1 Å (10.6 cm). Figures 7-1 and 6-7*a* also show that the major groove exposes the bases to the solvent, whereas in the minor groove only a small portion of the bases are accessible for interacting with molecules larger than water. This has profound implications for protein-DNA interactions.

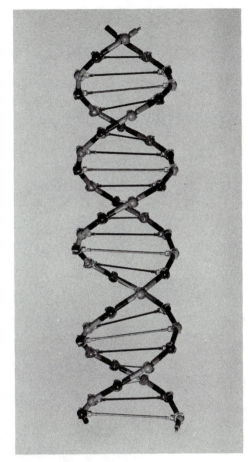

Figure 7-2
Model of B-DNA.

Table 7-1 Angle settings for B-DNA model

Chains	No.	Color	$\theta(°)$	$\tau(°)$
1 and 2	1 to 24	Red	31	199
		Green	31	199

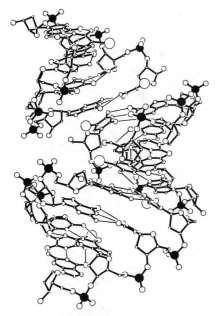

a

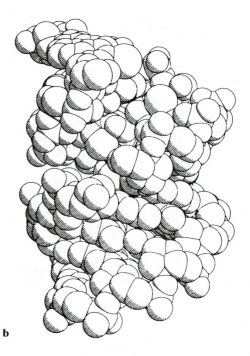

b

Figure 7-3
Skeletal (**a**) and space-filling (**b**) repre-
sentations of A-DNA. Note the propeller twist
of the base pairs in **a**. (From Dickerson RE,
Drew HR, Connor BN, et al: *Science* 1982;
216:475–485. Copyright 1982 American
Association for the Advancement of
Science.)

A-DNA Conformation

Conformation of the DNA molecule can be changed from
B-DNA to A-DNA in solution by raising the salt con-
centration. Under these conditions, the A form of DNA
appears to be more stable than the B form. A-DNA is
shown in Figure 7-3. The A form of DNA is structurally
homologous to double-stranded RNA. The minor groove
is very deep while the major groove is nearly flush with
the surface of the molecule. A good description of the
DNA molecule in the A conformation is that it is a twisted
ribbon.

The DNA molecule in the A conformation has eleven
base pairs per turn, and the bases are tilted about 20°
with respect to the long axes of the DNA molecule. The
base pairs are not coplanar but show 15° of propeller
twist, which improves base-stacking contacts. The phosphate groups still form a charged ridge on the outside of
the molecule, but now those along the minor groove are
close to one another. The biological role of A-DNA is
questionable, since physiological salt concentrations

rarely would allow A-DNA to be stable. However, it is possible that DNA may be induced to assume the A form upon association with specific proteins.

Building the A-DNA Model

The purpose of this exercise is to compare the structures of A-DNA and B-DNA. You will note that the major and minor grooves are changed but that the perpendicular twofold symmetry axes remain the same. (See p. 104.)

Building the A-DNA model is essentially the same as building B-DNA and takes the same amount of time. The scale for the A-DNA model is 0.67 cm/Å. The only changes are that the θ and τ angles are different (Table 7-2), and that the rods connecting the two strands are longer, 10.4 cm instead of 8.7 cm.

To build the model, alternate red and green units along each chain. Align and intertwine the two chains so that red 1 of chain 1 is opposite green 24 of chain 2. Clip on the twenty-four 10.4 cm rods with connectors and extend or compress the model so that the major groove is 11 cm and the minor groove is 4 cm. The rods are at an angle of about 70° to the helix axis. The two chains are antiparallel. Figure 7-4 shows the completed model.

On examination of the models and Figures 7-2 and 7-4 you can see that the rotation and translation that generate a single strand in both A- and B-DNA act on the unit of a single nucleotide base. The Z-DNA structure described later is different in this regard, since the rotation and translation are applied to two nucleotide base pairs for a single strand.

Having 11 nucleotides per turn instead of ten has an effect on the overall radius and length of the A-DNA molecule. This molecule of 24 nucleotides is 10%

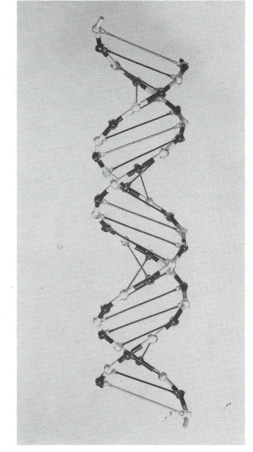

Figure 7-4
Model of A-DNA.

Table 7-2 Angle settings for A-DNA model

Chains	No.	Color	$\theta(°)$	$\tau(°)$
1 and 2	1 to 24	Red	29	195
		Green	29	195

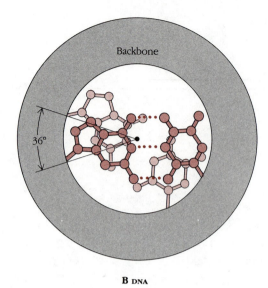

B DNA

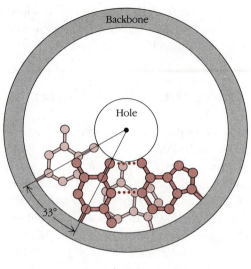

A DNA

Figure 7-5
When viewed along the helix axis, B-DNA is solid, while in A-DNA a central hole is apparent.

shorter; it is 45 Å long instead of 50 Å long (B-DNA) and has a radius of 12 Å instead of 10 Å for B-DNA (compare the two backbone models). As seen in Figure 7-5, the A conformation shows a central "hole" along the long axis. A particularly striking feature of A-DNA is that the minor groove is very different from that seen in the B conformation.

In both the A- and B-DNA models, the antiparallel arrangement of the two strands and the twofold rotational symmetry axes are apparent. Two different twofold rotational axes can be seen in the completed DNA models. One is along the line drawn from the phosphorus atom (the ball, not the rod) and intersects the perpendicular DNA helix axis. The second twofold axis is also perpendicular to the helix axis but passes through the mid-point of the support rod (in the center of the Watson-Crick base pairs). Find these axes on the models. Note that these symmetry axes do not apply when considering the chemical identity of the bases.

Z-DNA Conformation

The Z conformation of DNA was discovered recently by Rich and associates at the Massachusetts Institute of Technology. This form appears in concentrated salt solu-

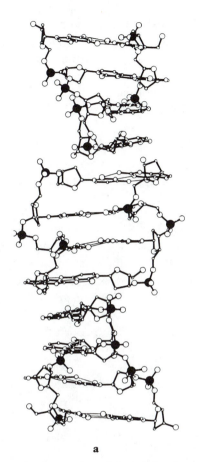

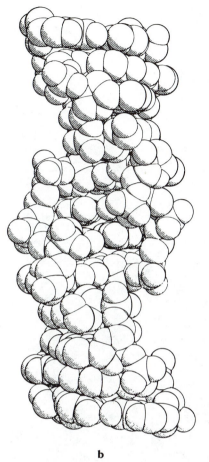

a   b

Figure 7-6
Z-DNA in skeletal (**a**) and space-filling (**b**) representations. Z-DNA is a left-handed double helix with a double nucleotide pair repeat. Note that the base-pair propeller twist is minor. (From Dickerson RE, Drew HR, Connor BN, et al: *Science* 1982; 216:475–485. Copyright 1982 American Association for the Advancement of Science.)

tions, but only when the bases G and C strictly alternate. The G-C (or more generally, purine-pyrimidine) alternation allows a completely novel type of DNA structure (Figure 7-6). The Z-DNA molecule is a left-handed helix instead of the right-handed helix just described for the A- and B-DNA molecules.

Unlike the sinuous **S** curve seen in the backbone of A- and B-DNA, the Z-DNA backbone is a dramatic zigzag. This unique backbone structure arises because of a rotation of the base about the base to deoxyribose bond. There is no major groove in the Z structure, but the minor groove is cavernously deep. Furthermore, the tilt of the base pairs is slightly different, about 10°. There is almost no propeller twist of the base pair in Z-DNA. The bases are virtually coplanar, in contrast to A- and B-DNA. Most importantly, the helical operation that generates the double-stranded Z-DNA operates on two base pairs (four nucleotides). The Z-DNA conformation is favored by the base-stacking of G-C pairs and underscores the importance of packing interactions as a major determinant of macromolecular conformation. A complete discussion of Z-DNA conformation can be found listed in the Suggested Readings at the end of this chapter.

Building the Z-DNA Model

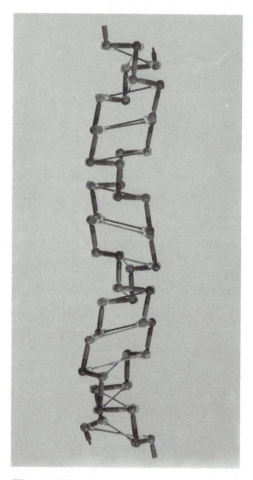

Figure 7-7
Model of Z-DNA.

The Z-DNA model shows the dramatically different conformation that a double-stranded DNA molecule may have. A major point is that the translational and rotational helix operator works on a pair of nucleotides per strand.

This model is difficult to build and takes about 90 minutes. Use two sets of 24 nucleotide units and the angles in Table 7-3 to construct a DNA backbone model in the Z conformation. Figure 7-7 shows a completed model, which is built by following these instructions.

Cut 24 rods 5.6 cm long. Remove the smaller half of all 48 connectors and assemble onto the rods. Align the clips on the rods such that they open in opposite directions; that is, one cavity is facing right, the other left (Figures 7-8 and 3-18). The two chains should be aligned and intertwined antiparallel such that the red unit 1 of chain 1 is opposite the green unit 24 of chain 2. Setting the first rod is difficult because the clips offer no support. The cavities do not sit under the ball but rather on the side. You will get the feel of it as you progress. With the rod in place (use the thumb and index fingers on one hand to main-

Table 7-3 Angle settings for Z-DNA model

Chains	No.	Color	$\theta(°)$	$\tau(°)$
1 and 2	1 to 24	Red	96	357
		Green	83	301

tain a slight pressure on the rod) brush a small amount of glue on both clips. Let this set. Place the next rod in place and proceed. The rods are at an angle of about 80° to the helix axis. The first two rods are difficult to set. However, once these are set, the remainder are easy!

Temporary (not glued) spacer rods must be inserted to allow for the major and minor groove separation. Attach these as follows:

| Minor groove | Red-red | Cut a 4.8 cm rod |
| Major groove | Green-green | Cut a 7.6 cm rod |

These rods should be moved down the structure as the base pair rods are set (Figure 7-8). It is critical that you use these spacer rods to set the first pair of cross rods. Failure to do so may result in poor alignment of the two chains.

It is not necessary to use the major and minor groove spacer rods after you have set the first pair of permanent cross rods, but you should check the alignment and spacing of the major and minor grooves as you proceed along the chain. The position of the spacer rods used to set the first two cross rods is as follows:

1. *Short spacer rod 1* (red to red): Place a 4.8 cm rod (with intact connectors) between position 3 on chain 1 and position 17 of chain 2.

2. *Short spacer rod 2* (red to red): Place a 4.8 cm rod between positions 5 on chain 1 and position 15 on chain 2.

3. *Long spacer rod 1* (green to green): Place a 7.6 cm rod between position 8 on chain 1 and position 22 on chain 2.

4. *Long spacer rod 2* (green to green): Place a 7.6 cm rod between position 10 on chain 1 and position 20 on chain 2.

With these four spacer rods in place, manipulate the structure such that the outline of both chains forms a hexagonal pattern when viewed down the helix axis.

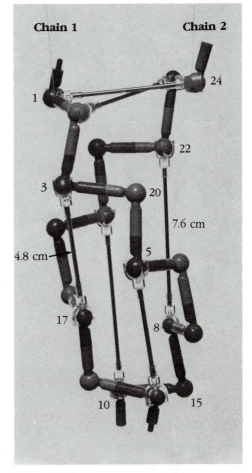

Figure 7-8
Construction of the model of Z-DNA is aided by the use of support rods, which measure the major and minor grooves.

The rods (5.6 cm) form crossed pairs as follows:

Chain 1		Chain 2	
Red	1	Green	24
Green	2	Red	23
Red	3	Green	22
Green	4	Red	21

As you can see, the zigzag backbone is quite obvious. The scale for Z-DNA is 0.53 cm/Å. Note the positions of the two twofold symmetry axes perpendicular to the long axis of the Z-DNA model. These are similar to what is seen in A- and B-DNA, since the symmetry axes are a general property of two-start, antiparallel helices. The structural differences between A, B, and Z forms of DNA are summarized in Table 7-4. Compare your three models (keeping the scale differences in mind) to see the dimensions of the helices.

Table 7-4 Structural differences of A, B, and Z forms of DNA

Characteristic	B-DNA	A-DNA	Z-DNA
Handedness	Right	Right	Left
Repeating unit	Base pair	Base pair	Two base pairs
Base pairs per turn	10	11	12
Degrees per base pair	36	33	30
Rise of helix per base pair	3.4 Å	2.9 Å	G-C 3.5 Å C-G 4.1 Å
Length of repeat unit	34 Å	32 Å	46 Å
Helix diameter	21 Å	23 Å	19 Å

The physiological role of Z-DNA is not known. Long stretches of alternating purine-pyrimidine bases exist in cellular DNA. It is not clear whether these regions of DNA assume the Z conformation, or if they do, what conditions cause or allow them to become Z-DNA.

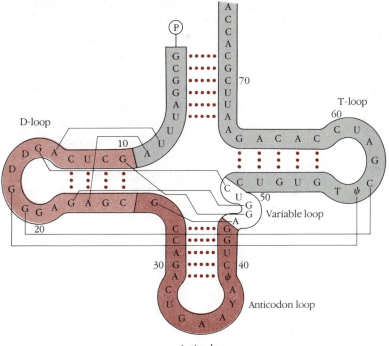

Figure 7-9
Nucleotide sequence and major features of
the base-base interactions in yeast
phenylalanine tRNA. There are four stems in
the molecule formed by double helices.
Additional bases interact through hydrogen
bonding to form the three-dimensional
molecule. These tertiary interactions are
shown by the long solid lines. The
anticodon stem and loop are colored, the D
stem and loop are lightly colored, and the
variable stem and loop are white. The T
stem and loop and amino acid acceptor
stem are shaded.

The tRNA Molecule

A tRNA molecule consists of a single strand of nucleic
acid 75 to 90 bases long. X-ray structures have been
determined for five different tRNA molecules and all
show the same general features.

The tRNA molecule is L-shaped. Four segments of
double helix in the molecule interact with each other as
shown in Figure 7-9. Double-helical RNA cannot assume
a B-DNA–like conformation because the 2′ hydroxyl
(absent in DNA) would collide with the phosphodiester
backbone on the successive residue. The four regions of
double-helical structure—the acceptor arm, T (Ψ) C
arm, anticodon arm, and D arm—are arranged in the two
arms of the L as shown in Figure 7-10.

The structure of tRNA is much more extended than that
observed for globular proteins of similar relative molecular weight. Both Watson-Crick hydrogen-bonding
and base-stacking interactions contribute to the final

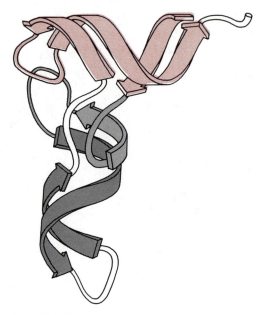

Figure 7-10
Backbone structure of a tRNA. The shading is the same as in Figure 7-9. (Courtesy Jane Richardson.)

tRNA structure. Additionally, distant regions of the cloverleaf come together in three-dimensional space to form so-called "tertiary base pairs." Figure 7-11 shows some of the tertiary interactions (defined in Figure 7-9) between the D loop, TΨC loop, and the variable loop. This structure involves three different strand segments from several regions of the molecule.

Other unusual features of the structure include numerous interactions between the bases and the phosphodiester backbone. The ribose 2'-hydroxyl groups can form hydrogen bonds and are thus also important in determining tertiary structure (Figure 7-11).

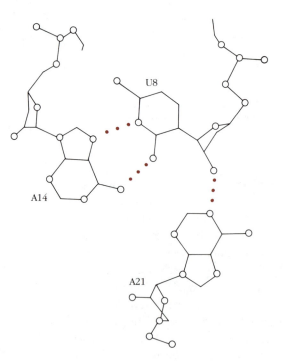

Figure 7-11
Some of the ribose 2'-OH groups can form tertiary interactions as seen here in the D helix.

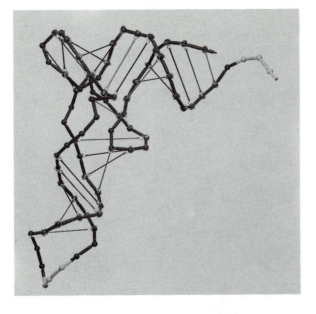

Figure 7-12
Model of yeast phenylalanine tRNA.

Building the Yeast Phenylalanine tRNA Model

The purpose of this exercise is to examine the beautiful three-dimensional structure of tRNA. The single-stranded nucleic acid molecule forms four stems that fold together into the characteristic L-shaped tRNA. The relationship between structure and function in tRNA is a mystery, since the anticodon and amino acid acceptor (CCA) are on opposite ends of the "L."

To build a model of the yeast phenylalanine tRNA molecule, it is convenient to make slight modifications in the positions of some of the phosphate groups. Not all phosphate groups are equidistant in the tRNA molecule. Thus to build a model of tRNA using model units of fixed length, the phosphates must be adjusted to make the best approximation to the structure. In the real yeast phenylalanine tRNA structure, the phosphate-phosphate distances are mostly 5.9 Å, but they range from 5 to 6.8 Å in the extremes.

Table 7-5   Angle settings for tRNA model

No.	Base	θ	τ	Color	Comments
1	pG	0	180	Green	5′ end
2	C	28	208	Green	
3	G	33	199	Green	
4	G	25	242	Green	
5	A	48	152	Green	
6	U	23	163	Green	
7	U	26	357	Green	
8	U	40	346	Green	
9	A	43	133	Green	
10	m²G	107	163	Green	
11	C	23	180	Green	
12	U	51	178	Green	
13	C	59	346	Green	
14	A	22	54	Green	
15	G	45	26	Green	
16	D	49	11	Green	
17	D	87	86	Green	
18	G	105	273	Green	
19	G	27	275	Green	
20	G	10	18	Green	
21	A	95	111	Green	
22	G	26	32	Green	
23	A	37	119	Green	
24	G	27	262	Green	
25	C	45	164	Green	
26	m²G	22	234	Green	
27	C	61	234	Green	
28	C	20	164	Green	
29	A	53	79	Green	
30	G	18	317	Green	
31	A	68	252	Green	
32	Cm	15	162	Green	
33	U	54	161	Green	
34	Gm	84	116	Red	Anticodon
35	A	34	182	Red	Anticodon
36	A	73	6	Red	Anticodon
37	Y	32	38	Green	
38	A	67	199	Green	
39	ψ	30	244	Green	

The tRNA model can be built in about 2 hours from 76 nucleotide units, with the CCA amino acid acceptor stem and the anticodon bases built in different colors from the remainder of the backbone. The scale factor for tRNA is 0.64 cm/Å and the virtual bond length between phos-

Table 7-5   continued

No.	Base	θ	τ	Color	Comments
40	mC	50	162	Green	
41	U	16	232	Green	
42	G	43	144	Green	
43	G	14	232	Green	
44	A	34	216	Green	
45	G	19	214	Green	
46	m^7G	28	202	Green	
47	U	67	48	Green	
48	C	85	106	Green	
49	mC	131	82	Green	
50	U	50	187	Green	
51	G	23	199	Green	
52	U	22	178	Green	
53	G	44	229	Green	
54	T	17	202	Green	
55	ψ	45	159	Green	
56	C	98	122	Green	
57	G	32	119	Green	
58	mA	38	280	Green	
59	U	79	93	Green	
60	C	44	295	Green	
61	C	56	226	Green	
62	A	35	193	Green	
63	C	54	101	Green	
64	A	7	277	Green	
65	G	58	262	Green	
66	A	2	127	Green	
67	A	35	213	Green	
68	U	31	181	Green	
69	U	41	213	Green	
70	C	29	208	Green	
71	G	15	150	Green	
72	C	46	236	Green	
73	A	40	175	Green	
74	C	22	196	Red	Amino acid acceptor
75	C	43	180	Red	
76	A	0	180	Red	3′ end

phate atoms is 5.9 Å on average. The model requires the following: 70 green units, 6 red units, 27 rods and 54 clips. Build segments of ten nucleotides each, and set the θ and τ angles as shown in Table 7-5. Cut and assemble the base pair connecting rods and link the nucleotide

Table 7-6 Connecting rods

Positions		Cut (cm)	Comments
72	1	10.0	
71	2	9.8	
70	3	10.0	
69	4	9.6	
68	5	9.6	Nonclassical Watson-Crick base pair
67	6	9.6	
66	7	10.0	
65	49	10.0	
64	50	9.9	
63	51	10.2	Cut small cup half off connector for 63
62	52	9.5	
61	53	9.8	
43	27	9.3	
42	28	9.0	
41	29	9.0	
40	30	8.7	Nonclassical Watson-Crick base pair
39	31	8.9	
25	10	9.3	Cut small cup half off connector for 10
24	11	9.3	Nonclassical Watson-Crick base pair
23	12	10.4	
22	13	10.5	Nonclassical Watson-Crick base pair
8	14	4.6	
48	15	10.0	Cut small cup half off connector for 48
18	55	8.8	
19	56	8.0	
26	44	9.8	
54	58	5.3	Cut small cup half off connector for 58

phosphates according to Table 7-6. A model of yeast phenylalanine tRNA is shown in Figure 7-12.

Note that the tRNA backbone is a right-handed helix whose structure is mainly determined by base-stacking and the hydrogen bonding of Watson-Crick base pairs. The double-helical segments of tRNA appear similar to the A-DNA helix. Notice also that the amino acid acceptor or CCA stem (which is linked to the cognate amino acid) and the anticodon loop (which reads the code for the cognate amino acid) are on opposite ends of the molecule, separated by 85 Å. This is presumably important for tRNA function, since all known tRNAs are similar in structure.

Summary

Double-helical DNA may exist in many forms inside the cell. Three forms—A-DNA, B-DNA, and Z-DNA—have been studied by x-ray crystallographic analysis. B-DNA is the best known form and is an antiparallel, right-handed double helix with a repeat unit of ten nucleotides, per turn. A-DNA is like B-DNA but is shorter and thicker and resembles a twisted ribbon. The Z form of DNA is a left-handed helix whose structure is different from other DNA forms that have been examined. Z-DNA is an extended antiparallel double helix. All three forms have two twofold rotational axes of symmetry. Some characteristics of the binding of cellular regulatory proteins to DNA is probably reflected in this symmetry. The different forms of DNA expose different portions of the bases to the environment or, more specifically, DNA-binding proteins. It is possible that, in performing their biological functions, regulatory proteins may recognize or induce particular conformations of DNA.

Transfer RNA is a nucleic acid whose tertiary structure as well as primary sequence plays an important part in its function. Since all known tRNAs have similar structures, the L-shape and spacing of anticodon and amino acid acceptor stem must be of particular biological importance.

Suggested Readings

DNA Conformations

Dickerson RE: The DNA helix and how it is read. *Sci Am* 1983; 249:94–111. Review of the modern findings in the field of DNA structure that is well illustrated, well written, and simple to understand. Presentation of the important structural details of base roll and propeller twist that have important sequence-dependent variation in helical DNA structure.

Dickerson RE, Drew HR, Connor BN, et al: The anatomy of A, B, and Z DNA. *Science* 1982; 216:475–485. Excellent comparison of these three conformers of DNA.

Zimmerman SB: The three-dimensional structure of DNA. *Ann Rev Biochem* 1982; 51:395–427. Excellent review of most aspects of the structures of the several conformations of DNA that have been identified. The solution properties of DNA are also discussed in a structural context.

Z-DNA

Wang AHJ, Quigley GJ, Kolpak FJ, et al: Molecular structure of a left-handed double helical DNA fragment at atomic resolution. *Nature* 1982; 282:680–686. Discussion of Z-DNA.

Nordheim A, Pardue ML, Lafer EM, et al: Anitbodies to left-handed Z-DNA bind to interband regions of drosophila polytene chromosomes. *Nature* 1981; 294:417–422. Account of possible localization and identification of Z-DNA.

tRNA

Rich A, Kim SH: The three-dimensional structure of transfer RNA. *Sci Am* 1978; 238:52–62. Presentation of the x-ray structure of a tRNA.

Solvent-accessible surface representation of part of the flavodoxin molecule. On the left are α helices: one is viewed end-on; the other, obliquely. They are packed against the five strands of the parallel β sheet. The curvature of the sheet is apparent from the strand rotations. The helices covering the right side of the sheet and all side chains are omitted for clarity. (See Figure 5-19.)

PART III

MODELS OF PROTEINS

In Part III we present instructions for constructing models of insulin, the *cro* repressor, and the enzyme lysozyme. These particular proteins were chosen because detailed knowledge of their structures is available, they are not cumbersome to construct, and they illustrate many aspects of protein structure and its relationship to function.

The insulin model illustrates the joining of two polypeptide chains by disulfide bridges and the principle of linking segments of secondary structure to construct a protein. The residues that are probably involved in binding to the insulin receptor are also discussed. There are other polypeptide hormones whose amino acid sequences are very similar to insulin's. These insulin homologs probably have structures much like insulin, and are an excellent example of the fact that a three-dimensional structure will accommodate different amino acid sequences. This point will be discussed later.

The *cro* repressor is discussed because it is different from typical globular proteins, presumably because it has the highly specialized function of binding a very large ligand, DNA. The *cro* repressor is representative of the larger class of proteins that bind DNA and regulate the expression of its genetic information. The interactions of *cro* repressor with DNA will be discussed. Like insulin, the *cro* molecule also has structural homologs.

The enzyme lysozyme has approximately the same number of amino acids as the *cro* dimer but has a strikingly different structure. Lysozyme is a typical small globular protein about twice the size of insulin. Details of the interaction of lysozyme with its substrate, the oligosaccharide that makes up bacterial cell walls, is well understood. You will construct models of lysozyme and its substrate and examine the interaction between the two molecules.

8

Insulin

The physiology of the insulin response has been understood for decades. The three-dimensional structure of the insulin molecule has been elucidated relatively recently, but the insulin receptor remains an elusive target for structural and biochemical study. Because insulin is a small polypeptide, functional aspects are particularly obvious in protein structure. Three insulin-like hormones are known. It is tempting to speculate on the functional role of the differences and similarities between these homologs and insulin.

Insulin: A Polypeptide Hormone

Insulin is a small polypeptide hormone that is responsible for regulating the level of blood sugar. The fundamental role of insulin in metabolism has led to an enormous amount of study of its structure and function. Insulin increases the rate at which fatty acids, proteins, and glycogen are synthesized. One of insulin's major functions is to promote the conversion of glucose in the bloodstream into liver glycogen, which is stored for later use. Glycogen storage granules in a liver cell are seen in Figure 8-1. Insulin serves as a signal that tells cells to

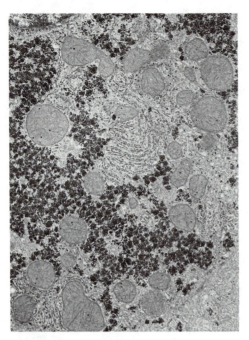

Figure 8-1

Electron micrograph of a well-fed liver cell. Dark particles are glycogen, which makes up 6.8% of the cell mass. The individual small particles that make up the glycogen rosettes are 300 Å in diameter and are coated with the enzymes involved in building and degrading glycogen. (×19.4.) (From Striffler JS, Cardell EL, Cardell RR: *Am J Anat* 1981; 160:363–379.)

store energy. After one eats a meal, blood glucose levels rise. This triggers the pancreas to secrete insulin into the bloodstream (Figure 8-2). Without insulin, the body is unable to remove glucose from the blood and store it in the form of glycogen, and a condition called *hyperglycemia* results. The best known example of pathological hyperglycemia is the disease diabetes mellitus, which is characterized by an abnormally high blood sugar level.

Because of insulin's central role in regulating cellular processes, it has been the subject of numerous studies by biochemists and physiologists. In fact, the first amino acid sequence of any protein was determined for insulin in 1953 by Frederick Sanger. He was awarded the Nobel prize for this work. He showed that the molecule is composed of two chains; the A chain has 21 residues, and the B chain has 30 residues. These chains are bound together by two disulfide bonds (Figure 8-3). Insulin is made as a

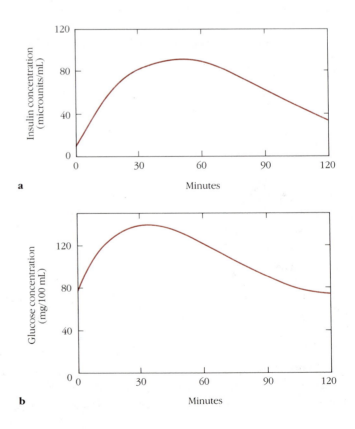

Figure 8-2

a, The level of insulin in the blood rises after eating and falls back to the resting level after about 2 hours. **b**, The level of glucose in the blood shows a similar pattern to that seen for insulin. Glucose is removed from the bloodstream and is stored as glycogen. Insulin regulates the body's response to an increased blood glucose level.

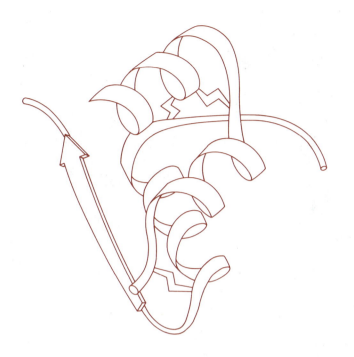

Figure 8-3
Schematic drawing of insulin showing the
two chains and disulfide bonds (represented
by zigzags) joining them. (Courtesy Jane
Richardson.)

single polypeptide precursor, proinsulin, which is post-
translationally modified to yield the final product. The
modification is the removal of an internal protein segment of 33 amino acids: the C peptide. Figure 8-4 is a
schematic illustration of proinsulin and its three
peptides.

Insulin is a protein signal that is carried in the bloodstream and binds to a specific receptor on the surface of
all cells that respond to it. This binding process and a
number of subsequent steps result in the alteration of the
physiological function of responsive cells. The association of insulin and its receptor is extraordinarily
tight. Indeed at 10^{-10} M there are few stronger dissociation constants between biological macromolecules.
Tight binding is required because small amounts of insulin are secreted by the pancreas and must find their
way to other tissues in the body (such as muscle and
liver).

Insulin structure has been studied using x-ray crystallography (see Blundell et al. in the Suggested Readings
at the end of the chapter). The insulin crystals that were
worked on in this laboratory are probably not too dis-
similar from crystals found naturally in pancreatic storage

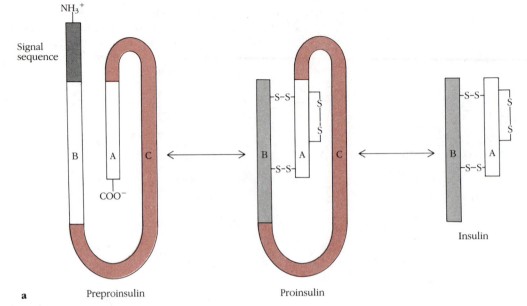

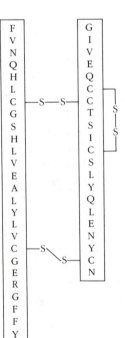

a

Preproinsulin

Proinsulin

Insulin

Figure 8-4

a, Schematic diagram of the proinsulin molecule. It is derived from preproinsulin by proteolytic hydrolysis of the peptide bond linking the signal peptide to proinsulin. The cellular secretion of the molecule is directed by the signal sequence peptide. **b**, The primary sequence of the polypeptide is written on the ribbon. (**a** from Stryer L: *Biochemistry*. San Francisco, W.H. Freeman & Co., 1981, p. 848.)

b

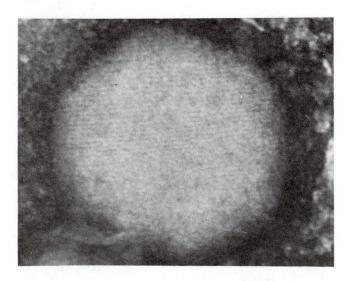

Figure 8-5
Electron micrograph of a portion of a pancreas cell showing a crystalline array of insulin molecules in a storage granule. Note the parallel horizontal rows. They are separated by 50 Å. Presumably the physiological conformation of insulin is similar to that observed in the manmade insulin crystals used in x-ray crystallographic experiments. (×400,000.) (Courtesy Paul Lacy.)

granules (Figure 8-5). In the following section, a model of insulin will be built, and some structural features that give insulin its biochemical and physiological properties will be discussed. In addition, insulin will be compared with a class of other hormones that probably have a similar structure based on comparison of the amino acid sequences of these different proteins.

Building the Insulin α-Carbon Model

This model will take approximately one hour to build. You will construct two chains: Chain A has 21 units and Chain B, 30 units. Set θ and τ angles as listed in Tables 8-1 and 8-2. Cut three support rods to the lengths 3.4, 4.5, and 3.1 cm, and attach connectors. These should link amino acids A3-V and B28-P; A17-E and B18-V; and A10-I and B4-Q, respectively.

Now link the cysteines with three support rods cut to 3.5 cm to form three disulfide bonds. Add two connectors to each rod to make completed rod assemblies. Link the following pairs of α-carbons: A6-A11, A7-B7, and A20-B19.

Table 8-1   Angle settings for insulin monomer model: insulin A chain

No.	Color	Amino Acid	$\theta(°)$	$\tau(°)$
1	Red	Gly G	0	180
2	Green	Ile I	80	230
3	Green	Val V	89	222
4	Red	Glu E	83	231
5	Red	Gln Q	86	232
6	Green	Cys C	89	217
7	Green	Cys C	78	218
8	Green	Thr T	84	179
9	Red	Ser S	53	107
10	Green	Ile I	44	15
11	Green	Cys C	41	51
12	Red	Ser S	49	62
13	Green	Leu L	94	242
14	Green	Tyr Y	89	228
15	Red	Gln Q	89	237
16	Green	Leu L	90	252
17	Red	Glu E	92	252
18	Red	Asn N	94	238
19	Green	Tyr Y	82	285
20	Green	Cys C	62	287
21	Red	Asn N	0	180

Figure 8-6
Model of insulin.

Table 8-2 Angle settings for insulin monomer model: insulin B chain

No.	Color	Amino Acid	$\theta(°)$	$\tau(°)$
1	Red*	Phe F	0	180
2	Green	Val V	45	82
3	Red	Asn N	61	26
4	Red	Gln Q	32	39
5	Red	His H	77	355
6	Green	Leu L	57	351
7	Green	Cys C	47	159
8	Red	Gly G	61	176
9	Red	Ser S	91	252
10	Red	His H	91	233
11	Green	Leu L	89	233
12	Green	Val V	89	232
13	Red	Glu E	89	230
14	Green	Ala A	89	230
15	Green	Leu L	91	232
16	Green	Tyr Y	90	226
17	Green	Leu L	84	223
18	Green	Val V	82	211
19	Green	Cys C	76	344
20	Red	Gly G	68	180
21	Red	Glu E	86	252
22	Red	Arg R	97	29
23	Red	Gly G	58	50
24	Green	Phe F	30	67
25	Green	Phe F	60	23
26	Green	Tyr Y	74	351
27	Green	Thr T	65	358
28	Green	Pro P	57	87
29	Red	Lys K	64	150
30	Red*	Ala A	0	180

*These amino acids are hydrophilic because they are at the termini and consequently charged.

The model should now have the approximately correct conformation. Align the model, according to Figure 8-6, into the correct three-dimensional conformation. Fix all flexible connectors to the balls using a small amount of glue.

Figure 8-7
α-Carbon drawing of insulin.

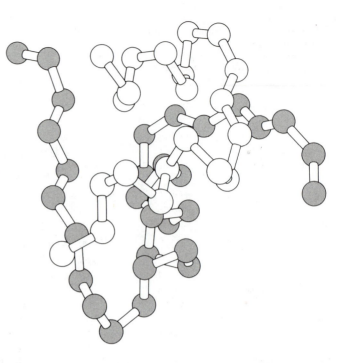

Functional Aspects of Insulin Structure

Since the scale of the model is 1 cm/Å, the model is 100 million times larger than the actual molecule. (A baseball magnified to the same degree would be the size of the earth!) The molecule is approximately $30 \times 30 \times 20$ Å. Insulin is one of the smallest proteins known but has a typical globular structure. The protein has two layers, consisting of the A and B chains. See Figures 8-7 and 8-8 for a backbone and all-atom representation of insulin. Together these chains form a molecule of a single domain. This is a rare example of a single domain consisting of separate polypeptide chains. This phenomenon can be explained by the fact that insulin is made and folds as a precursor, proinsulin. The intervening peptide is then removed from the already formed insulin molecule. Thus, this domain may be thought of as consisting of a single polypeptide. Figures 8-9 to 8-12 show the side chains and backbone conformation for the separated A and B chains.

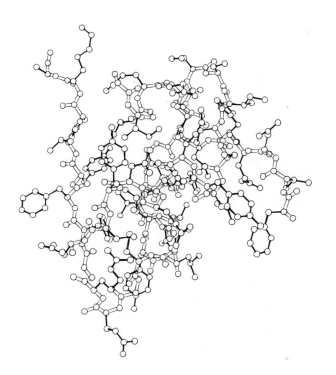

Figure 8-8
All-atom representation of insulin. The positions of the α-carbons are the same as in the ribbon drawing (Figure 8-7).

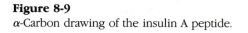

Figure 8-9
α-Carbon drawing of the insulin A peptide.

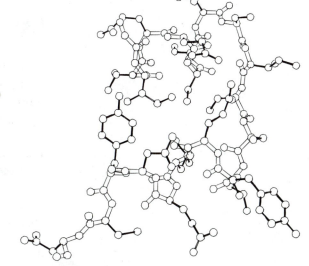

Figure 8-10
All-atom representation of the insulin A peptide.

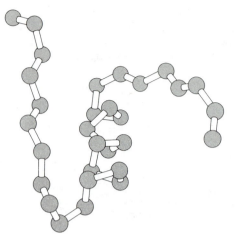

Figure 8-11
α-Carbon drawing of the insulin B peptide.

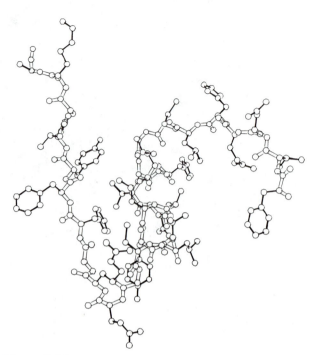

Figure 8-12
All-atom representation of the insulin B peptide.

The two chains are held together covalently by two of insulin's three disulfide bonds. Figure 8-13 is a simple illustration of a disulfide bond. The third disulfide linkage is intrachain and stabilizes the A chain. Association of the interior hydrophobic amino acids and hydrogen bonds between side chains contribute to insulin's stability. Note that the inner core of the model is completely green (hydrophobic) and that the surface is mostly red (hydophilic). Remember that the N and C termini of the chains are charged even though three of the side chains are hydrophobic. Find the N terminus of the A chain and the C terminus of the B chain. Note that these are on the surface and near each other in the model. The C peptide in proinsulin connects these two ends.

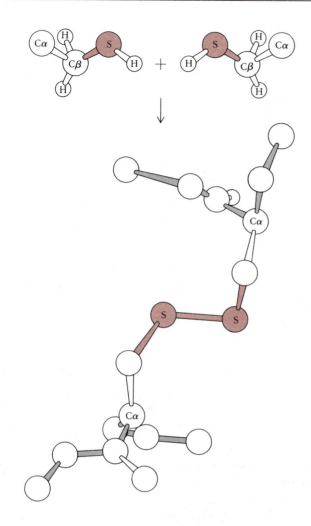

Figure 8-13
Disulfide bonds form between the side chains of two cysteine residues. Two —SH (thiol) groups are oxidized to form one —S—S— (disulfide) group.

The interior of the protein is not open as the backbone model suggests but is filled with the atoms of the buried hydrophobic side chains. Figure 8-14 is a space-filling drawing that more accurately represents the interior packing of the protein. In Figure 8-14, a plane transects the molecule at about amino acids A11, B7, B22, and B26. The atoms are drawn with their effective van der Waals radii.

The A and B chains of insulin are covalently attached by disulfide bonds, but van der Waals interactions and hydrogen bonds (for example, A4-E to B19-K) contribute significantly to insulin's stability. The interior of insulin is

Figure 8-14
Space-filling drawing of the interior of
insulin. Note that the atoms are close-
packed. This represents a cross-sectional
plane through the center of the molecule.

close-packed even though it is composed of two individual polypeptide chains. The features of close atomic
packing and noncovalent bonding are responsible for the
stability of most protein-protein interactions and are not
unique to insulin.

Helices and Sheets: Secondary Structure in Insulin

There is only one modest β-sheet region in the protein.
Look at residues 10 to 13 of the A chain and 2 to 5 of the B
chain. These form a short two-stranded antiparallel β
sheet. Note that the strands are separated by 4.7 Å and
that along the sheet the amino acids repeat at about 6.8 Å,
the usual distance separating alternating amino acids in β
structure. The β strands in this sheet are not arranged

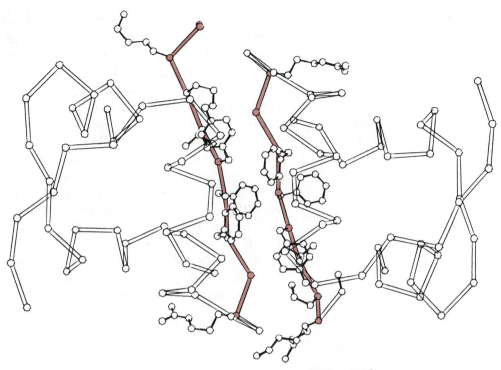

Figure 8-15

α-Carbon and all-atom representations of the insulin dimer show the β sheet formed by the two strands at the C termini of the two B chains.

with the typical 10° to 15° angle between them that is seen in most β sheets. Instead they intersect at an angle of 45°. Thus only about three hydrogen bonds can be formed between side chains of the two strands, as compared to five or six in the typical case. The A and B chains in the insulin structure are stabilized further by inter-chain disulfide bonds.

The B peptide has a second β strand, amino acids 23 to 30. This strand has no apparent mate. It is unusual to find a long unpaired β strand in a protein. It might be expected that this extended strand would be part of a β sheet. This is, in fact, the case. Insulin normally exists as a dimer in solution with a dissociation constant of about 10^{-5} M. As shown in Figure 8-15, when two insulin monomers associate, they form another two-stranded sheet from the unpaired β strands, amino acids B23 to B30 of each B chain.

There are three helices in the molecule: from A1 to A9, A12 to A20, and B9 to B20. These have the dimensions of typical α helices, with a backbone diameter of 5 Å or 10 Å if side chains are included. The rise per residue is 1.5 Å. However, the helices contain two turns of 3_{10} helix with about three residues per turn, at A15 to A20 (note that this is the end of an α helix). If you look down the helix axis, you will see the differences between the α and 3_{10} helices. The 3_{10} helix is more extended, using seven residues (A14 to A20) to traverse 11 Å as compared to eight residues for the α helix (B11 to B18). Note that the helices are all right-handed and that they generally have a green (hydrophobic) side and a red (hydrophilic) side.

The insulin monomer has four chain reversals or turns. There are two in the A chain between the first α helix and the short β strand (C7 to I10; type II) and the short β strand and the second α helix (S12 to Q15; type III). In the B chain the turns are at C7 to H10 (type II) and C19 to R22 (type II). Note that three of these turns have cysteines that form the disulfide bonds.

Insulin-Receptor Binding

The insulin receptor (Figure 8-16), which has been purified from plasma membrane, is composed of two different-sized subunits. The molecular weight of the insulin-binding subunit is 135,000 daltons whereas that of insulin is 5700 daltons. Insulin is about 25 times smaller than its receptor. The 135,000 dalton subunit is associated in the membrane with a 90,000 dalton protein that has phosphotransferase activity and does not bind insulin. The two subunits are made from a single polypeptide precursor that is cut by a specific protease.

By comparing insulin from different species and chemically modified insulin analogs, a group of amino acid side chains and molecular surfaces presumably involved in receptor binding have been identified. Some of these residues are hydrophobic, for example, B24-F, B25-F, B26-Y, B12-V, and B16-Y. Notice that these are located on the surface and are green in the model. The hydrophobic surface facing solvent is probably involved in mating with a similar and complementary surface on

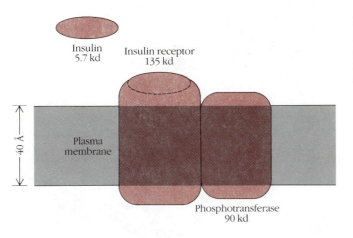

Figure 8-16
Schematic diagram showing the relative sizes of insulin, the two subunits of the insulin receptor, and the plasma membrane of the cell.

the receptor. It is *very unusual* to find so many hydrophobic amino acids clustered on a protein's surface, and this area might have been proposed as the receptor-binding region by simply examining the model.

Certain hydrophilic amino acids are thought to make hydrogen bonds with the receptor. Generally nonessential side chains (especially if they are on the surface) vary in different species, but those that are required for function or stability are conserved (usually identical but possibly replaced by amino acids with similar chemical characteristics). For insulin, conserved amino acids include A1-G (which is charged because it is at the N terminus), A5-Q, A19-Y, A21-N, and A23-R. Note that the amino acids implicated in receptor binding are scattered across the surface of the insulin model. This suggests that the receptor covers much of the surface of insulin. See Figure 8-17 for a space-filling representation of insulin's surface.

Proinsulin and Hormones with Homologous Sequences

Since a mixture of proinsulin and insulin can be crystallized in crystals that are virtually identical to crystals of insulin, proinsulin is presumed to have a similar shape

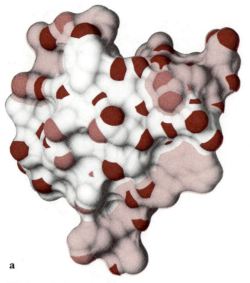

a

b

Figure 8-17
Space-filling drawings of insulin. Refer to
Figure 2-9 for the coloring scheme of the
atoms. In addition, the B chain is indicated
by the light shading. **a**, Hydrophilic face of
the molecule showing the A chain.
b, Hydrophobic face (B chain), which is
presumably involved in receptor binding.

and conformation as insulin. The position of the C peptide is predicted to be as is shown in Figure 8-18.

The other hormones in Figure 8-18 are thought to have structures similar to insulin, since they share considerable sequence homology. Relaxin acts to relax smooth muscle (the birth canal) whereas insulin-like growth factor (IGF) is responsible for proper growth during development. The Pygmy tribes of Kenya have recently been shown to be genetically deficient in IGF.

The sequences of these other hormones are given in Figure 8-19. You can convince yourself that the proposed structural similarities are plausible by tracing the sequences on the insulin model. Observe that all amino acid substitutions on the interior are conservative (for example, isoleucine for valine), whereas nonconserved amino acids are on the surface and presumably reflect differences in receptor structures.

Summary

Of the score of characterized polypeptide hormones, insulin is one of the best understood. Its central role in regulation of glycogen and fatty acid metabolism has

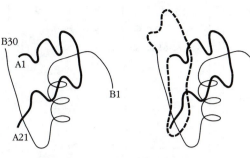

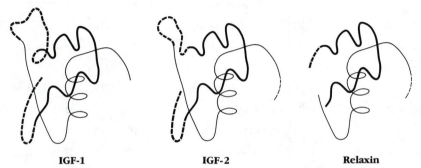

Figure 8-18

Schematic drawing of the backbones of insulin, proinsulin, and the homologous hormones IGF and relaxin. (Redrawn from Blundell TL, Humbel RE: *Nature* 1980; 287:781–787. Copyright © 1980 Macmillan Journals Ltd.)

been thoroughly studied in an attempt to understand the nature of the disease diabetes mellitus. Insulin is secreted as a single chain of 84 amino acids that is activated by precise removal of the central C peptide of 33 amino acids. The remaining A and B chains are covalently linked with disulfide bonds and form a structure that is mostly α helical with two short β strands in the B chain — in every sense a typical small globular protein. A short 3_{10} helix and standard turns are also found in the structure. Insulin dimerizes by forming an intersubunit antiparallel two-strand β structure.

The pattern of conserved amino acids and the hydrophobic nature of the surface of the insulin molecules indicate the region of the molecule that interacts with the insulin receptor in the cell's plasma membrane.

The insulin structure can accommodate alternate side chains at many positions that are at the surface of the molecule. This structure also is predicted for other hormones such as IGF1, IGF2, and relaxin. This observation suggests an evolutionary relationship among these hormones and their receptors.

A Chains

	-2	-1	1	2	3	4	5	6	7	8	9	10	11	12	13	14	15	16	17	18	19	20	21	22	23	24	25	26	27	28	29	30
Insulin																																
BOVINE	—	—	G	I	V	E	Q	C	C	A	S	V	C	S	L	Y	Q	L	E	N	Y	C	N	—	—	—	—	—	—	—	—	—
HUMAN	—	—	G	I	V	E	Q	C	C	T	S	I	C	S	L	Y	Q	L	E	N	Y	C	N	—	—	—	—	—	—	—	—	—
RAT 1	—	—	G	I	V	D	Q	C	C	T	S	I	C	S	L	Y	Q	L	E	N	Y	C	N	—	—	—	—	—	—	—	—	—
RAT 2	—	—	G	I	V	E	Q	C	C	T	S	I	C	S	L	Y	Q	L	E	N	Y	C	N	—	—	—	—	—	—	—	—	—
GUINEA PIG	—	—	G	I	V	E	Q	C	C	T	G	T	C	T	R	H	L	Q	—	N	Y	C	N	—	—	—	—	—	—	—	—	—
CASIRAGUA	—	—	G	I	V	E	Q	C	C	T	N	I	C	S	R	N	Q	L	L	T	Y	C	N	—	—	—	—	—	—	—	—	—
COYPU	—	—	G	I	V	E	Q	C	C	T	N	I	C	S	R	N	Q	L	M	S	Y	C	N	D	—	—	—	—	—	—	—	—
HAGFISH	—	—	G	I	V	E	Q	C	C	H	K	R	C	S	I	Y	N	L	Q	N	Y	C	N	D	—	—	—	—	—	—	—	—
Insulin-like growth factor																																
IGF 1	—	—	G	I	V	D	E	C	C	F	R	S	C	D	L	R	R	L	D	M	Y	C	A	P	L	K	P	A	K	S	A	—
IGF 2	—	—	G	I	V	E	E	C	C	F	R	S	C	D	L	A	L	L	D	T	Y	C	A	T	—	P	A	K	S	E	—	—
Relaxin																																
PORCINE	R	M	T	L	S	E	K	C	C	E	V	G	C	I	R	K	D	I	A	R	L	C	—	—	—	—	—	—	—	—	—	—

B Chains

	-2	-1	1	2	3	4	5	6	7	8	9	10	11	12	13	14	15	16	17	18	19	20	21	22	23	24	25	26	27	28	29	30	31
Insulin																																	
BOVINE	—	—	F	V	N	Q	H	L	C	G	S	H	L	V	E	A	L	Y	L	V	C	G	E	R	G	F	F	Y	T	P	K	A	—
HUMAN	—	—	F	V	N	Q	H	L	C	G	S	H	L	V	E	A	L	Y	L	V	C	G	E	R	G	F	F	Y	T	P	K	T	—
RAT 1	—	—	F	V	K	Q	H	L	C	G	P	H	L	V	E	A	L	Y	L	V	C	G	E	R	G	F	F	Y	T	P	K	S	—
RAT 2	—	—	F	V	K	Q	H	L	C	G	S	H	L	V	E	A	L	Y	L	V	C	G	E	R	G	F	F	Y	T	P	K	S	—
GUINEA PIG	—	—	F	V	S	Q	H	L	C	G	S	N	V	E	T	L	Y	S	V	C	Q	D	R	H	R	P	S	E	—	—	—	—	—
CASIRAGUA	—	—	Y	V	G	Q	R	L	C	G	S	Q	L	Y	D	D	T	L	Y	S	V	C	K	H	R	Y	R	P	S	E	—	—	—
COYPU	—	—	Y	V	S	Q	R	L	C	G	S	Q	L	Y	D	D	T	L	Y	S	V	C	R	H	R	V	R	G	P	T	K	—	—
HAGFISH	—	—	R	T	T	G	H	L	C	G	K	D	L	V	N	A	L	Y	I	A	C	G	V	R	G	F	Y	D	P	T	K	M	—
Insulin-like growth factor																																	
IGF 1	—	—	G	P	E	T	L	C	G	A	E	L	V	D	A	L	Q	F	V	C	G	D	R	G	F	Y	F	N	K	P	T	—	—
IGF 2	A	Y	R	P	S	E	T	L	C	G	G	E	L	V	D	T	L	Q	F	V	C	G	D	R	G	F	Y	F	S	R	P	A	—
Relaxin																																	
PORCINE	S	—	—	—	—	—	—	—	—	A	C	G	R	E	L	V	R	L	W	V	E	I	C	G	V	S	—	—	—	—	—	—	—

C Chains

	1	2	3	4	5	6	7	8	9	10	11	12	13	14	15	16	17	18	19	20	21	22	23	24	25	26	27	28	29	30	31	32	33	34	35
Insulin																																			
BOVINE	R	R	—	—	E	V	E	G	P	Q	V	E	L	A	G	G	P	G	A	G	G	L	—	E	G	P	P	Q	—	—	—	—	—	K	R
HUMAN	R	R	E	A	E	D	L	Q	V	G	Q	V	E	L	G	G	G	P	G	A	G	S	L	Q	P	L	A	L	E	G	S	L	Q	K	R
RAT 1	R	R	E	V	E	D	P	Q	V	P	Q	L	E	L	G	G	G	P	G	A	G	D	L	Q	T	L	A	L	E	V	A	R	Q	K	R
RAT 2	R	R	E	V	E	D	P	Q	V	A	Q	L	E	L	G	G	G	P	G	A	G	D	L	Q	T	L	A	L	E	V	A	R	Q	K	R
GUINEA PIG	X	X	E	L	E	D	P	Q	V	E	Q	T	E	L	G	W	G	I	G	—	G	P	L	—	Q	—	Q	A	L	Q	—	—	—	X	X
Insulin-like growth factor																																			
IGF 1	—	—	—	—	—	—	—	—	—	—	—	—	G	Y	G	S	S	S	R	—	—	—	R	—	—	—	—	A	P	Q	T	—	—	—	—
IGF 2	—	—	—	—	—	—	—	—	—	—	—	—	S	R	V	S	R	R	S	R	—	—	—	—	—	—	—	—	—	—	—	—	—	—	—

Figure 8-19 These protein sequences are homologous. The corresponding tertiary structures are presumed to be similar.

Suggested Readings

Insulin Function

Krahl ME: Endocrine function of the pancreas. *Ann Rev Physiol* 1974; 36:331–360. Discussion of the physiological role of insulin.

Insulin Receptor

Czech MP: Structure and functional homologies in the receptors for insulin and the insulin-like growth factors. *Cell* 1982; 31:8–10. Brief review of insulin receptor and its possible homologs.

Hedo JA, Kahn CR, Hiyashi M, et al: Biosynthesis and glycosylation of the insulin receptor. Evidence for a single precursor of the major subunits. *J Biol Chem* 1983; 258:10,020–10,026. Evidence that the insulin receptor is made as a single polypeptide that is cut to generate the two subunits with different functions.

Roth RA, Cassell DJ: Insulin receptor: evidence that it is a protein kinase. *Science* 1983; 219:299–301.

Insulin Structure

Blundell TL, Cutfield JF, Cutfield SM, et al: Atomic positions in rhombohedral two-zinc insulin crystals. *Nature* 1971; 231:506–511. Discussion of the structure of insulin.

Pullen RA, Lindsay DG, Wood SP, et al: Receptor-binding region of insulin. *Nature* 1976; 259:369–373. Account of receptor binding of insulin.

Insulin Homologs

Blundell TL, Humbel RE: Hormone families: pancreatic hormones and homologous growth factors. *Nature* 1980; 287:781–787. Discussion of structure-function relationships in the insulin family.

9

cro Repressor

For a DNA-binding protein to have a specific effect on gene expression, there must be sequence-specific recognition between protein and DNA. A particularly good example of sequence-specific binding is seen in the lysis-lysogeny decision of the bacteriophage λ. The *cro* and *cI* repressors bind to the same DNA sequences with different affinities. These two proteins have strikingly similar secondary structures that are involved in interaction with the DNA molecule. Clearly the binding preferences must lie in subtle tertiary structural differences between the protein-DNA complexes.

cro Repressor: A modulator of Gene Expression in Phage λ

The *cro* repressor is a small protein (66 amino acids) that is coded for by the phage λ, which infects the bacterium *E. coli*. The λ life cycle in *E. coli* can proceed two ways. In the lytic cycle, the cell produces progeny phage and eventually lyses. In the lysogenic cycle the free λ phage becomes covalently integrated into the *E. coli* genome and is no longer expressed.

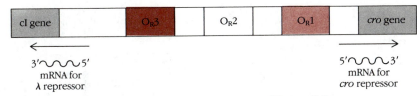

3′ ∿∿∿ 5′
mRNA for
λ repressor

5′ ∿∿∿ 3′
mRNA for
cro repressor

Figure 9-1
The λ phage O_R DNA and the adjacent genes, *cI* and *cro*.

The λ genome contains a central regulatory region consisting of two operators, O_R and O_L, which bind phage-encoded DNA regulatory proteins. The phage λ genes are arranged about these operators so that transcription in one direction yields gene products predominately used in the lytic cycle (phage-replication enzymes and structural proteins). Transcription in the other direction generates proteins required for the lysogenic cycle.

It is obvious that a subtle switching process is involved in determining which life cycle λ will take. By prohibiting transcription in one direction while allowing transcription in the other, the lysis-lysogeny decision is made. To understand gene regulation in this system, one need only to examine O_R, because O_L is analogous. (See Figure 9-1 for a schematic representation of the O_R region.) λ DNA codes for two major repressor proteins, *cro* and the λ repressor (cI gene), which bind at O_R. O_R contains three 18 base-pair regions (O_R1, O_R2, and O_R3) of homologous sequence, which have been shown by DNA protection experiments to be the exact sites of repressor binding. *Cro* protein and the gene product *cI* compete for binding to these sites, thus determining the direction of transcription from O_R. These 18 base-pair regions are similar but not identical, and in fact, the different repressors prefer certain sites over others. *Cro* protein binds to O_R3, physically blocking the DNA so that RNA polymerase cannot bind and initiate transcription of the *cI* gene. Thus *cro* protein allows rightward transcription of itself and prevents leftward transcription of *cI*. Similarly, the *cI* gene product binds at O_R1 and O_R2, blocking the site where RNA polymerase must bind to transcribe *cro*. This prevents transcription of *cro*. Interestingly, *cI* binding at O_R2 (presumably through interactions of *cI* protein and the RNA polymerase) enhances

Figure 9-2

This is an inverted repeat in a DNA sequence. Note the symmetry in the sequence.

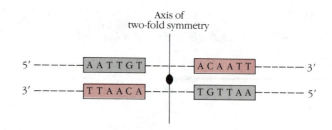

RNA polymerase binding in the O_R3 region so that transcription of the *cI* gene to the left is stimulated. The sequences of 18 base pairs in the operator segments are each imperfectly symmetrical (in sequence) about a twofold rotational axis perpendicular to the long axis of the DNA. This sort of symmetry in DNA sequence is known as an *inverted repeat* (Figure 9-2). The operator O_R1 is shown in Figure 9-3.

Anderson and co-workers (1981) have determined the crystal structure of the *cro* repressor and have found the monomer to be relatively simple. The protein contains three α helices and three β strands with a disordered C terminus (Figure 9-4). The functional *cro* structure that binds DNA is a dimer with a twofold symmetry axis. One monomer may be superimposed on the second monomer of the dimer by a 180° rotation. The twofold symmetry axis of this dimer is aligned with the twofold symmetry axis of the DNA, as you will see. The dimer presumably binds to DNA with its twofold axis at the center of the inverted repeat or twofold axis of symmetry of the operator.

Chemical protection experiments have proven that the *cro* repressor protects seven separate guanines in the 17-base pair operator fragment from methylation at the N7 position. Because the N7 position is exposed only in the major groove, this suggests that binding of the *cro* repressor to DNA blocks the major groove to access by chemicals. It has also been shown that ethylation of the phosphate backbone inhibits *cro* binding. The DNA protection results provide a model of how the *cro* repressor interacts with the operator. This postulated model is consistent with those deduced for other protein-DNA interactions, for example, RNA polymerase and promoters.

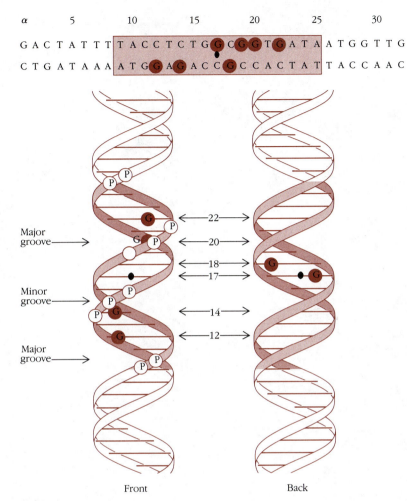

Figure 9-3

Sequence of operator O_R1 and a cartoon of λ O_R1 DNA region. This is a representation of the front and back of 16 base pairs of B-DNA. *P* corresponds to backbone phosphate groups that are believed (from chemical protection experiments) to be physically near the λ repressor when it is bound to DNA. *G*s are guanine bases that are protected from chemical attack when λ repressor is bound to DNA. The protein is believed to bind to the front of the DNA as is depicted in this figure. (From Pabo CO, Krovatin W, Jeffrey A, Sauer RT: *Nature* 1982; 298:441–443. Reprinted by permission from copyright © Macmillan Journals Ltd.)

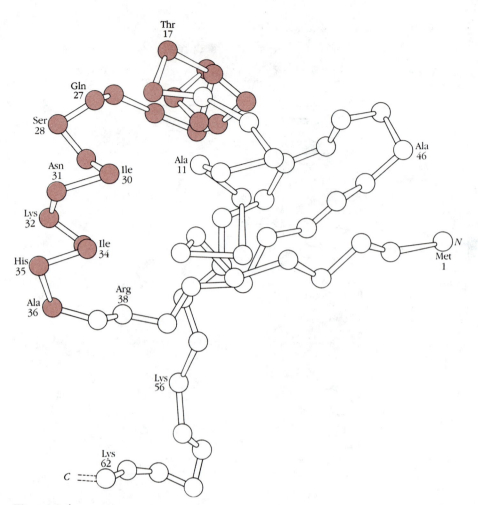

Figure 9-4
The backbone structure of the *cro* monomer has three α
helices in succession. In the monomer these interact with a β
sheet composed of three strands. The positions of several of
the amino acids are indicated. The disordered C terminus is
indicated by the dashed line. (From Anderson WF, Ohlendorf
DH, Takeda Y, Matthews BW: *Nature* 1981; 290:754–758.
Reprinted by permission from copyright © Macmillan
Journals Ltd.)

Figure 9-5
Three *cro* repressor dimer molecules bound to the three operators of the λ O_R region. Note that the *cro* dimer binds only one face of the DNA molecule and that the C termini appear to wrap around it. (Courtesy Wayne Anderson.)

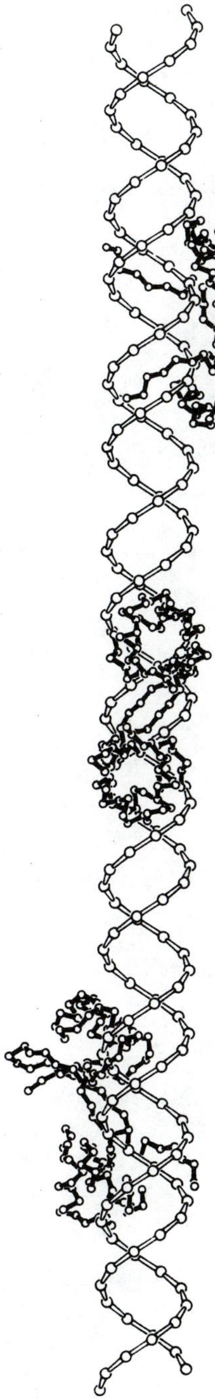

The protein seems to bind predominantly to one side of the DNA double helix (Figures 9-3 and 9-5). *Cro* repressor is thought to protect bases in the major groove by aligning an α helix in the groove. The dimer thus protects two regions of the DNA using an α helix from each monomer.

Building the *cro*-Dimer Model

The *cro* repressor monomer model can be built using Table 9-1, which lists the bend and torsional angles. Construct two monomer models. Once these models are completely assembled, they should be joined with connecting rods.

The length and positions of the monomer links are as follows:

No. 1	No. 2	Length (cm)
5	42	3.3
8	34	4.1
10	50	7.5*
11	27	8.4*
23	51	4.1
40	55	4.3

Because this is a dimer, you must link the two monomers together. The following dimer links are the key for setting the symmetry axis correctly:

Monomer 1	Monomer 2	Length (cm)
25	59	4.4
59	25	4.4
54	57	4.1
57	54	4.1

*Temporary; do not glue.

Table 9-1 Angle settings for *cro*-repressor monomer model*

No.	Color	Amino Acid	$\theta(°)$	$\tau(°)$
1	Red	Met M	0	180
2	Red	Glu E	43	33
3	Red	Gln Q	67	14
4	Red	Arg R	66	353
5	Green	Ile I	53	73
6	Green	Thr T	64	69
7	Green	Leu L	88	236
8	Red	Lys K	80	223
9	Red	Asp D	90	237
10	Green	Tyr Y	94	228
11	Green	Ala A	88	235
12	Green	Met M	87	218
13	Red	Arg R	73	206
14	Green	Phe F	80	63
15	Red	Gly G	54	94
16	Red	Gln Q	89	235
17	Green	Thr T	85	228
18	Red	Lys K	95	227
19	Green	Thr T	77	235
20	Green	Ala A	95	231
21	Red	Lys K	82	228
22	Red	Asp D	89	219
23	Green	Leu L	88	51
24	Red	Gly G	86	300
25	Green	Val V	51	83
26	Green	Tyr Y	67	67
27	Red	Gln Q	86	246
28	Red	Ser S	96	239
29	Green	Ala A	93	231
30	Green	Ile I	85	226
31	Red	Asn N	81	228
32	Red	Lys K	86	229
33	Green	Ala A	87	220
34	Green	Ile I	94	233
35	Red	His H	84	227
36	Green	Ala A	97	56
37	Red	Gly G	87	359
38	Red	Arg R	63	66
39	Red	Lys K	80	336
40	Green	Ile I	60	18
41	Green	Phe F	53	38
42	Green	Leu L	58	23
43	Green	Thr T	60	13

Table 9-1   (continued)

No.	Color	Amino Acid	$\theta(°)$	$\tau(°)$
44	Green	Ile I	68	9
45	Red	Asn N	61	90
46	Green	Ala A	88	252
47	Red	Asp D	102	65
48	Red	Gly G	93	357
49	Red	Ser S	68	38
50	Green	Val V	81	352
51	Green	Tyr Y	39	23
52	Green	Ala A	27	338
53	Red	Glu E	41	53
54	Red	Glu E	63	28
55	Green	Val V	50	351
56	Red	Lys K	46	82
57	Green	Pro P	53	62
58	Green	Phe F	59	221
59	Green	Pro P*	99	357
60	Red	Ser S	90	24
61	Red	Asn N	35	321
62	Red	Lys K	51	12
63	Red	Lys K	41	351
64	Green	Thr T	69	39
65	Green	Thr T	65	213
66	Red	Ala A	0	180

*We have ignored the correct distance of 2.7 Å between *cis*-Pro and the next amino acid.

The completed model appears in Figure 9-6, and an α-carbon drawing is shown in Figure 9-7. These figures show that an extended (and out-of-register) six-stranded β sheet is formed by the linkage of the individual β strands from the two monomers. This sheet is unusual because only two or three strands are aligned at a time as discussed next.

cro Repressor Structure

The assembled *cro* repressor can now be compared with an assembled B-DNA model. Remember that the scales of the two models are different. The B-DNA model is two-

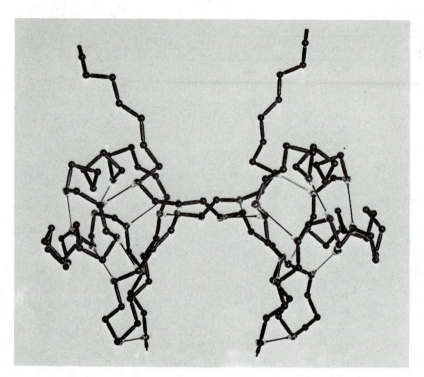

Figure 9-6
Model of *cro* dimer. Note that the dimer has
an extensive six-stranded β sheet.

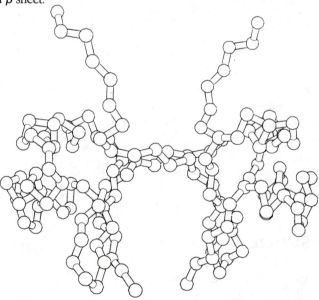

Figure 9-7
α-Carbon drawing of the *cro* dimer.

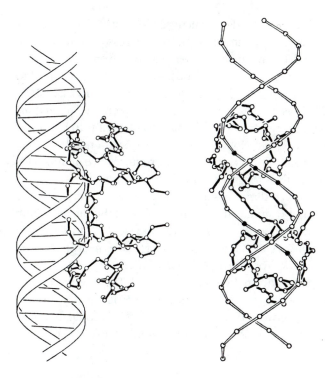

Figure 9-8
The repressor-DNA interaction shows the
twofold symmetry axes to be superimposed.
(From Anderson WF, Ohlendorf DH, Takeda
Y, Matthews BW: *Nature* 1981; 290:754–758.
Reprinted by permission from copyright ©
1982 Macmillan Journals Ltd.)

thirds the size it should be to be compared to the protein
model. The twofold axes of the DNA molecule and *cro*
repressor dimer can be superimposed to demonstrate
the interaction depicted in Figure 9-8. The α helices
labelled *gln 27* to *ala 36* in Figure 9-4 fit neatly into the
major grooves on the front face of the B-DNA molecule as
seen in Figure 9-3. Notice that the structure of the *cro*
repressor is unlike that of other globular proteins (for
example, insulin), since the α-helical portions of the
protein protrude from the molecule's surface. The three
α helices are held out so that the third one, $\alpha 3$, can insert
into the major groove.

Note that the arrangement of the hydrophobic amino
acids (green) on the three helices with those of the β
strands form a hydrophobic core. Helix $\alpha 3$ is two sided,
as is typical for α helices. One face is hydrophobic
(green) and forms part of the hydrophobic core, while
the second face is hydrophilic (red) and associates with
the hydrophilic surface of the major groove of DNA.

The six secondary structural elements (three α helices and three β strands) in the *cro* repressor monomer are linked by five turns. As you can see from your model, these turns are found at positions 6 to 9 (type III), 12 to 15 (type I), 34 to 37 (type I), 45 to 48 (type III), and 23 to 27. This last turn is the turn between the α_2 and α_3 helices and does not fall into any class of turn discussed in this book. Notice that the turn is extended and that reversal of the chain direction requires five amino acids rather than four as seen in other turn types.

The *cro* dimer has a very unusual structure. Notice the arch of β strands that form one face of the dimer. Within a monomer, β strands 1 (amino acids 1 to 7) and 2 (amino acids 41 to 46) form a stable hydrogen bond network. The third β strand (amino acids 48 to 59) is separated from the second β strand. The C termini of the two monomers (amino acids 55 to 59) form a hydrogen-bond network that links the monomers (compare *cro* to the insulin dimer). The role of amino acids 50 to 66 is not known.

Comparison with Other Repressors

You have observed that the *cro* repressor protein has two subunits related by a twofold axis symmetry. This twofold axis of symmetry is aligned with a twofold axis of symmetry on the DNA to form the DNA-protein complex. Each subunit of the protein contains an α helix, $\alpha 3$, which protrudes from the surface of the protein in such a way that the symmetrical α helices of the two subunits are parallel and separated by 34 Å. This is the exact separation distance between successive major grooves on right-handed DNA in the B form. A second feature of the $\alpha 3$ helix is that it is oriented at the correct angle to fit into the major groove. Thus the structure of *cro* repressor fits neatly against DNA, and it is possible that the pair of $\alpha 3$ helices specifically binds to the exposed major groove bases. This implies that recognition of the operator is partly accomplished by interaction of the specific sequence of side chains that emanate from the $\alpha 3$ helix with bases in the operator. The details of this interaction are not known, but the interface will be close-packed with specific hydrogen bonds between the side chains of the helix α_3 and the polar atoms of the bases that are exposed in the major groove.

It is unusual to see two contiguous α helices connected with the particular angle that is seen in the case of $\alpha2$ and $\alpha3$. Thus this tight turn structure may only be formed from a special arrangement of amino acids that are found at and near the joint between $\alpha2$ and $\alpha3$.

From comparison of the structure of the λ and *cro* repressors, it has been found that an $\alpha2\alpha3$-like arrangement is in λ repressor and that the amino acid sequences of the bend regions are homologous. The amino acid sequences are known for a number of other repressor molecules and it is interesting to determine whether they have the same type of general structure that you have observed for *cro* repressor and whether the other repressors bind to DNA in the same way that *cro* repressor does. Catabolite Activator Protein (CAP) is a DNA-regulatory protein that controls expression of many operons in *E. coli*. When the glucose level is normal or high, CAP is inactive and does not bind DNA. When the glucose level falls, cAMP is produced. CAP then binds cAMP and is activated to bind to DNA and turn on transcription of many operons involved in catabolism of other sugars. The CAP protein functions primarily as an activator of RNA polymerase and not as a repressor. The CAP structure has been determined at near atomic resolution by x-ray crystallography. It has been shown that 22 of the α-carbon atoms of the corresponding $\alpha2$ and $\alpha3$ helices align with *cro* repressor in three-dimensional space within an average error of 1.1 Å for each pair of amino acids. In fact, with the exception of repressor proteins, this two-helix motif has not been found in any other of the two hundred or so proteins for which atomic coordinates are available.

DNA-protein interaction is more complicated than $\alpha3$ helices fitting neatly into the major groove of B-DNA. Even though the $\alpha2$ and $\alpha3$ helices of CAP monomer align exactly with those of the *cro* monomer, the CAP *dimer* structure reveals a different spatial organization (tilt and orientation) of the pair of $\alpha3$ helices, which are 34 Å apart. It is not understood how CAP binds to DNA.

		Start helix α2								**Turn**			**Start Helix α3**									
	15				20						25					30					35		
λ *cro*	G	Q	T	K	*T*	A	K	D	*L*	*G*	*V*	Y	Q	S	A	*I*	K	A	*I*	H	A	G
λ cII	G	T	E	K	*T*	A	E	A	*V*	*G*	*V*	D	K	S	Q	*I*	S	R	*K*	R	D	W
434 *cro*	T	Q	T	E	*L*	A	T	K	*A*	*G*	*V*	K	Q	Q	S	*I*	Q	E	*L*	A	G	V
P22	R	Q	A	A	*L*	*G*	K	M	*V*	*G*	*V*	S	N	V	A	*I*	S	Q	*W*	E	R	S
cI	S	Q	E	S	V	A	D	K	*M*	*G*	*M*	G	Q	S	G	V	G	A	*F*	N	G	I
1ac	T	L	Y	D	V	A	E	Y	*A*	*G*	*V*	S	Y	Q	T	V	S	R	V	N	Q	A
gal	T	I	K	D	V	A	R	L	*A*	*G*	*V*	S	V	A	T	V	S	R	V	I	N	N
CAP	T	R	Q	E	*I*	*G*	Q	I	*V*	*G*	*C*	S	R	E	T	V	G	R	*I*	L	K	M

Figure 9-9
Amino acid sequences of bacterial and viral repressors and activators showing homology to the α2α3 two-helix motif of *cro* repressor. The sequence numbers are for the *cro* repressor only. Conserved amino acids are italicized.

Comparison of Repressor Sequences: α_2 and α_3 Helices

Figure 9-9 gives the amino acid sequences for *E. coli*, CAP, *gal*, and *lac*; phage λ, *cro*, *cI*, and *cII*; and phage P22 repressors. These sequences have been aligned on the basis of greatest homology. Six of the 22 amino acids in the α2 and α3 helices are identical for CAP and *lac* repressor, and furthermore, six others are conservative substitutions. The homology of the *cro* repressor sequence to these repressors is weaker. However, since the three-dimensional structures are known for the *cro* protein, the CAP activator, and *cI* repressors, the sequence alignment presented in Figure 9-9 is soundly based on structural data.

Certain residues are essential for formation of the structurally unique two-helix motif. These are obvious in the α2 and α3 segments of CAP shown in Figure 9-10. The absolutely conserved amino acids are involved in the interaction between the α2 and α3 helices or in formation of the extremely tight turn between them. The amino acids that are most variable in the sequence comparison tend to have polar side chains. This means they are on the surface and interact with either solvent or the DNA itself. The most important residues (or side chains), which are strictly conserved and are presumably important for determining the structure of these two helices, are (a) a glycine seen in Figure 9-10 at position 24 in the *cro* repressor; (b) a glycine or alanine at position 20 in

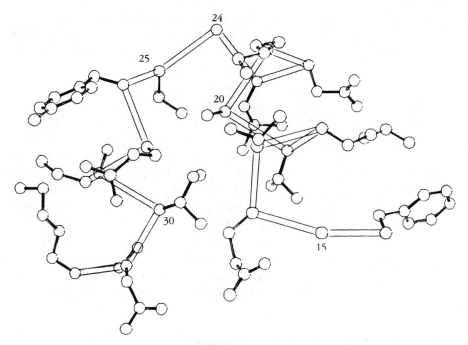

Figure 9-10
The hydrophobic interactions at the turn between α2 and α3 are apparent in this drawing of part of the *cro* protein. The numbers of the amino acids are for the *cro* protein only.

the *cro* repressor; and (c) a valine or isoleucine at position 30 in the *cro* repressor sequence. We also note that glycine at position 24 is always surrounded by two nonpolar amino acids. Glycine is appropriate at this position in the tight turn between the two helices. The ϕ and ψ angles that define this part of the backbone severely restrict the choice of amino acid that will occupy this position. In fact, the side chain of any amino acid other than glycine would sterically collide with the backbone and possibly perturb the two-helix motif.

Position 20 must be either an alanine or glycine, since the β-carbon atom points directly into an interior pocket with a very small volume. This tiny pocket is caused by folding the backbone into the tight turn between the two helices α2 and α3. This is obvious if one examines the α-carbon model of the *cro* repressor (Figure 9-10). Additionally, the valine or isoleucine that is conserved at position 30 is involved in a significant packing interaction between these two α helices. If you trace the α2 and α3 helix sequences on the *cro* α-carbon model, you clearly see that there are no residues involved in the interaction between the two helices which are not at least partially conserved. All these observations suggest that (a) the two-helix structural motif exists in many phage and bacterial DNA regulatory proteins, and (b) this two-helix motif is involved in specific DNA operator-sequence recognition. The specific interaction will vary for the several repressors.

By comparing the protein structure and the DNA structure, one may predict which side chains are important in forming the interaction. This can be done by mating a B-DNA model with the *cro* repressor. It appears that the interaction would start about three or four base pairs outside the center of twofold symmetry in the operator sequence and continues for an additional four or five base pairs on either side of this approximate twofold axis. There are about seven base pairs in the middle of the operator that do not interact with the $\alpha 3$ recognition helix. Other regions of the *cro* repressor are likely to interact with the center of the operator.

Summary

The *cro* repressor protein is one of several DNA-binding, regulatory proteins whose tertiary structures have been determined. The life cycle of the bacteriophage λ is modulated by the interaction of *cro* and *cI* proteins with phage DNA. A model for *cro* (and by analogy, other repressor proteins) binding to DNA has been proposed based on structural features that are common among *cro* and other DNA-binding proteins. Many of the proteins studied have two α helices, $\alpha 2$ and $\alpha 3$, which are linked by a tight turn that holds them in a particular conformation. In addition, these proteins bind to DNA as dimers that possess a twofold rotational symmetry axis. The model suggests that the repressor dimer binds to DNA with its twofold axis aligned with the center of an inverted repeat and that the two α_3 helices are aligned in the major groove of the DNA, permitting sequence-specific recognition between amino acid side chains and nucleotide bases. Presumably the structure of the $\alpha_2 \alpha_3$ turn and a few other hydrophobic amino acids are required for proper alignment of the α_3 helices, since these features are extremely well-conserved between *cro* and other repressor proteins.

In addition to the α_2 and α_3 helices, the *cro* monomer contains a third α helix (α_1) and three β strands. In the *cro* dimer, β strands form an out-of-register, six stranded β sheet. The *cro* dimer is an unusual protein; it is a long molecule containing two lobes that are separated by a thin bridge. This unique structure is used for binding the large ligand DNA at two binding sites separated by 34 Å.

Suggested Readings

Reviews

Furth ME: Repressor represses *cro* represses repressor re-presses *cro*. *Nature* 1980; 286:330–331. Short description of how the λ repressor and *cro* interact.

Johnson AD, Ptashne M, Pabo C: A genetic switch in a bacterial virus. *Sci Am* 1982; 247:128–140. Readable review of the λ repressor system.

Ptashne M, Jeffrey A, Johnson AD, et al: How the lambda re-pressor and *cro* work. *Cell* 1980; 19:1–11. Thorough descrip-tion of the interactions of the two repressors with operators 1, 2, and 3.

cro Structure

Anderson WF, Ohlendorf DH, Takeda Y, Mathews BW: Structure of the *cro* repressor from bacteriophage and its interaction with DNA. *Nature* 1981; 290:754–758. Discussion of *cro*-repressor structure and speculation that the paired $\alpha3$ helices are aligned in the major groove.

Ohlendorf DH, Anderson WF, Fisher RG, et al: The molecular basis of DNA-protein recognition inferred from the structure of *cro* repressor. *Nature* 1982; 298:718–723. Presentation of a proposed molecular interaction between the $\alpha3$ helix of *cro* and DNA.

cro Homologs

Matthews BW, Ohlendorf DH, Anderson WF, Takeda Y: Struc-ture of the DNA-binding region of *lac* repressor inferred from its homology with *cro* repressor. *Proc Natl Acad Sci USA* 1982; 79:1428–1432. Discussion of $\alpha2$ and $\alpha3$ helices in *lac* repressor and *cro*.

Steitz TA, Ohlendorf DH, McKay DB, et al: Structural similarity in the DNA-binding domains of catabolite gene activator and *cro* repressor proteins. *Proc Natl Acad Sci USA* 1982; 79:3097–3100. Discussion of $\alpha2$ and $\alpha3$ helices in CAP and *cro*.

10

Lysozyme: Enzyme-Substrate Complex

The final chapter of this text deals with the construction of a model of an enzyme and its substrate. You will begin by building a standard backbone model of the protein and finish the exercise by adding all-atom side chains to the active site and constructing an all-atom model of the substrate. The construction of the complete model and the subsequent discussion is complicated and thorough but is extremely rewarding.

The use of all-atom models in this context is particularly fruitful. As mentioned in Chapter 1, all-atom models tend to be very confusing visually. This problem is obviated by construction of only a small region of the protein in atomic detail. The all-atom models allow a sophisticated discussion of the atomic interactions of enzyme and substrate. One may examine the flexibility and variable positioning of the amino acid side chains and substrate, observing how they may fill space. A disadvantage of the enzyme substrate model described is that the protein backbone is not represented in atomic detail, and thus its interactions with the substrate may not be discussed.

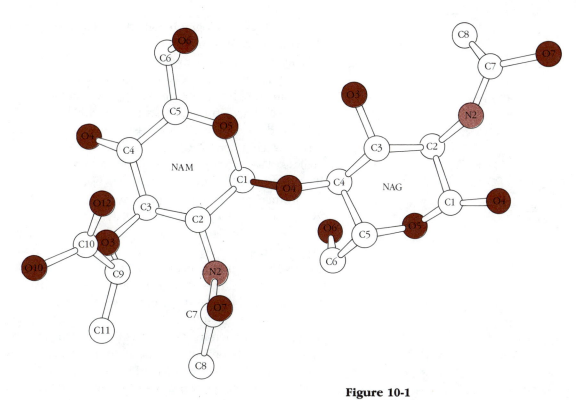

Figure 10-1
The sugars NAM and NAG joined by a β 1-4 linkage as they are found in the bacterial cell wall. In an α 1-4 linkage, the O4 oxygen would be below the plane of the page. The colored bonds show the β conformation of the C1 to O4 bond. The atomic numbering sequence is indicated.

Lysozyme and NAM-NAG-NAM

The enzyme lysozyme was first discovered in human tears by Fleming in 1922 on the basis of its remarkable antibacterial nature. Lysozyme acts on the oligosaccharides that compose bacterial cell walls. The cell wall of many bacteria is formed from an alternating polymer of the saccharides *N*-acetylmuramic acid (NAM) and *N*-acetylglucosamine (NAG) (Figure 10-1). These sugars are joined in a β 1-4 linkage, and the bond, between C-1 of NAM and O-4 of NAG is cleaved by the enzyme. The bond is cleaved in a hydrolysis reaction. Hydrolysis is a type of bond breaking in which a molecule of water, in the form of H^+ and OH^- ions, is donated to the cleavage products.

Lysozyme is an abundant and relatively small enzyme. Hen egg white lysozyme is a single polypeptide chain of 129 amino acids and a molecular weight of 14,600. In fact, the first x-ray crystallographic structure of an enzyme was determined for this molecule.

While x-ray crystallographic analysis of the structure of an enzyme is straightforward, study of an enzyme-substrate complex is not. In nature, the complex of an enzyme and its substrate exists only fleetingly, for 10^{-3} seconds or less. Thus it is impossible to study the structure of such a complex using standard x-ray diffraction analysis. The structure may be inferred from analysis of the enzyme and substrate analogs.

Substrate analogs are competitive inhibitors of enzymes because they bind to the active site but are not acted on by the enzyme. These inhibitors may be co-crystallized with enzymes or be added to enzyme crystals to determine the structure of the enzyme-substrate analog complex. This approach yields information about the position of the substrate in the active site, as well as detailed structural information about the enzyme and substrate. Frequently, the structure of the enzyme changes slightly on binding substrate, but rarely is any change in conformation induced in the substrate itself.

Hen egg white lysozyme is a model enzyme for study of molecular architecture. Lysozyme is small and fairly simple, and the structure of both the enzyme and the enzyme-substrate analog complex have been determined at 2.5 Å resolution. The model of the hen egg white lysozyme-substrate complex yields information about the structure of the protein and oligosaccharide and the interaction of the protein with its complicated ligand.

Building the Lysozyme-Substrate Complex Model

This model is more difficult to build than the *cro* dimer, which has nearly the same number of amino acids. It will take about 4 hours to complete the model. The model has 19 side chains and the lysozyme substrate, which is constructed from LabQuip all-atom models. LabQuip models have the same scale as *Blackwell Molecular Models* (1 cm/Å). The Blackwell connectors used in conjunction with support rods may be used to attach amino acid side chains made from LabQuip models to the α-carbon backbone. The center-to-center distance between the α- and β-carbons in the model is conveniently scaled to the correct α-β carbon distance found in proteins.

The advantages of using LabQuip models are severalfold. An all-atom representation of an important region of a protein may be generated easily using the backbone model. Once the proper orientation of the β-carbon is set (this is discussed later), a LabQuip side chain may be attached. Because there is free rotation about all single bonds in LabQuip models, the precise conformation of the side chains may be adjusted further. The use of all-atom models allows examination of specific noncovalent interactions between side chains and between side chain and substrate. (The required LabQuip parts listed in the materials list can be ordered directly from LabQuip or Blackwell Scientific Publications, Inc.)

Lysozyme is a globular protein and is approximately twice the size (in dimensions and molecular weight) of insulin, another globular protein you have studied. In contrast, lysozyme is almost exactly the same molecular weight as the *cro* dimer, yet its shape and dimensions are dramatically different. Lysozyme is a much more typical protein than *cro*, since lysozyme is compact and globular.

Building the Lysozyme Backbone Model

This model should be assembled in three parts. You should make the ten-unit segments, glue them together using the appropriate τ values, and add the support rods for each part. Connect the three parts using the two appropriate τ values given in Table 10-1. Use the support rods for the four disulfides, and the remaining five supports to completely stabilize the lysozyme model.

Table 10-1 Angle settings for the lysozyme model

No.	Color	Amino Acid	$\theta(°)$	$\tau(°)$
1	Red	Lys K	0	180
2	Green	Val V	72	33
3	Green	Phe F	53	64
4	Red	Gly G	64	67
5	Red	Arg R	90	216
6	Green	Cys C	82	238
7	Red	Glu E	88	224
8	Green	Leu L	87	236
9	Green	Ala A	94	229
10	Green	Ala A	87	226
11	Green	Ala A	81	232
12	Green	Met M	93	231
13	Red	Lys K	87	230
14	Red	Arg R	96	237
15	Red	His H	87	72
16	Red	Gly G	80	304
17	Green	Leu L	81	288
18	Red	Asp D	79	173
19	Red	Asn N	85	343
20	Green	Tyr Y	77	173
21	Red	Arg R	92	120
22	Red	Gly G	78	310
23	Green	Tyr Y	66	53
24	Red	Ser S	63	76
25	Green	Leu L	92	240
26	Red	Gly G	88	243
27	Red	Asn N	94	225
28	Green	Trp W	78	228
29	Green	Val V	93	232
30	Green	Cys C	84	228
31	Green	Ala A	91	230
32	Green	Ala A	84	235
33	Red	Lys K	86	214
34	Green	Phe F	86	233
35	Red	Glu E	83	170
36	Red	Ser S	68	66
37	Red	Asn N	91	114
38	Green	Phe F	91	313
39	Red	Asn N	68	13
40	Green	Thr T	86	246
41	Red	Gln Q	89	306
42	Green	Ala A	57	352

Continued

Table 10-1 (continued)

No.	Color	Amino Acid	$\theta(°)$	$\tau(°)$
43	Green	Thr T	46	5
44	Red	Asn N	49	43
45	Red	Arg R	60	27
46	Red	Asn N	54	73
47	Green	Thr T	86	244
48	Red	Asp D	94	72
49	Red	Gly G	83	303
50	Red	Ser S	62	38
51	Green	Thr T	49	52
52	Red	Asp D	59	21
53	Green	Tyr Y	39	168
54	Red	Gly G	40	139
55	Green	Ile I	92	216
56	Green	Leu L	84	53
57	Red	Gln Q	80	14
58	Green	Ile I	64	63
59	Red	Asn N	52	32
60	Red	Ser S	82	253
61	Red	Arg R	92	181
62	Green	Trp W	61	221
63	Green	Trp W	72	196
64	Green	Cys C	37	62
65	Red	Asn N	59	340
66	Red	Asp D	90	59
67	Red	Gly G	81	254
68	Red	Arg R	78	267
69	Green	Thr T	71	13
70	Green	Pro P	68	234
71	Red	Gly G	100	336
72	Red	Ser S	58	360
73	Red	Arg R	72	27
74	Red	Asn N	83	344
75	Green	Leu L	88	231
76	Green	Cys C	89	43
77	·Red	Asn N	89	301
78	Green	Ile I	49	62
79	Green	Pro P	77	59
80	Green	Cys C	88	249
81	Red	Ser S	92	245
82	Green	Ala A	91	244
83	Green	Leu L	91	239
84	Green	Leu L	78	286
85	Red	Ser S	58	74

Continued

Table 10-1 (continued)

No.	Color	Amino Acid	$\theta(°)$	$\tau(°)$
86	Red	Ser S	85	259
87	Red	Asp D	77	20
88	Green	Ile I	87	289
89	Green	Thr T	85	227
90	Green	Ala A	88	231
91	Red	Ser S	86	224
92	Green	Val V	89	233
93	Red	Asn N	90	218
94	Green	Cys C	85	238
95	Green	Ala A	93	240
96	Red	Lys K	90	226
97	Red	Lys K	86	234
98	Green	Ile I	89	223
99	Green	Val V	81	230
100	Red	Ser S	86	258
101	Red	Asp D	83	105
102	Red	Gly G	53	241
103	Red	Asp D	86	34
104	Red	Gly G	69	158
105	Green	Met M	88	271
106	Red	Asn N	99	261
107	Green	Ala A	97	250
108	Green	Trp W	71	17
109	Green	Val V	92	233
110	Green	Ala A	85	228
111	Green	Trp W	89	229
112	Red	Arg R	84	216
113	Red	Asn N	78	199
114	Red	Arg R	73	213
115	Green	Cys C	71	280
116	Red	Lys K	69	183
117	Red	Gly G	80	290
118	Green	Thr T	67	87
119	Red	Asp D	85	6
120	Green	Val V	87	266
121	Red	Gln Q	94	258
122	Green	Ala A	93	245
123	Green	Trp W	87	221
124	Green	Ile I	81	303
125	Red	Arg R	69	208
126	Red	Gly G	87	311
127	Green	Cys C	52	42
128	Red	Arg R	88	92
129	Red	Leu L	0	180

Part I: Amino acids 1 to 40 Using a Sharpie or VWR Lab Marker, mark amino acids 5, 10, 15, 20, 25, and so on. This step is critical because it enables you to identify amino acids in the middle of the completed protein model. Cut the support rods as follows:

Amino acids	Rod length (cm)
1 to 40	3.7
3 to 38	3.2
9 to 29	4.0

Snap the connector cups onto the appropriate α-carbons, rotating the cup on the rod to seat the cup properly. Do *not* glue the connector cups onto the α-carbons yet.

Part II: Amino acids 41 to 90 Mark every fifth amino acid. Then cut the support rods as follows:

Amino acids	Rod length (cm)
41 to 84	3.0 (trim small connector clip for 41)
42 to 54	3.4
45 to 51	2.2
50 to 69	3.5
52 to 59*	2.6
53 to 83	4.2
62 to 73*	2.7
64 to 80†	3.0

Do *not* glue the connector cups onto the α-carbons!

Part III: Amino acids 91 to 129 Mark every fifth amino acid and cut the support rods as follows:

Amino Acids	Rod length (cm)
98 to 107*	5.3
105 to 111	4.7 (cut small clip off cup for 111)

Do *not* glue the support rods to the α-carbons!

---

*Temporary rods to be removed later. Mark these special rods with a small tag made of tape for their easy identification later.
†Disulfide bond.

Figure 10-2
Lysozyme model (without side chains or the substrate).

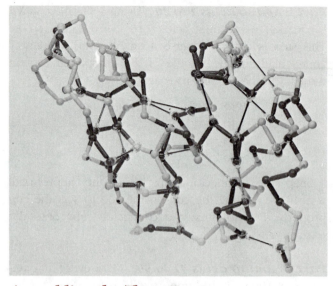

Assembling the Three Parts

Glue part I to part II using the τ angle for amino acid 40. Set the model down carefully, and check the τ angle once the model has dried to the touch (about 15 minutes). Now add supports to connect part I to part II. Cut the rod lengths as follows:

Amino acids	Rod length (cm)
32 to 55	3.8
35 to 57*	4.3

Do *not* glue the support rods to the α-carbons!

Prepare a support rod to connect α-carbons 30 and 115. This rod is cut to 4.2 cm long, and the final connector rod assembly is 6.2 cm long (this is a disulfide linkage between two cysteines). This support rod greatly helps to stabilize the model when you glue the third part, amino acids 91 to 129, to the rest of the model. Glue part III onto amino acid 90 using the τ setting for amino acid 90. Now add the support rod for amino acids 30 and 115. Again, do *not* glue the support rod; simply use it to stabilize the model. Let the glue for the τ angle for amino acid 90 dry. Then add the following support rods:

---

*Temporary support rod to be removed later. Mark with tape to remove later.

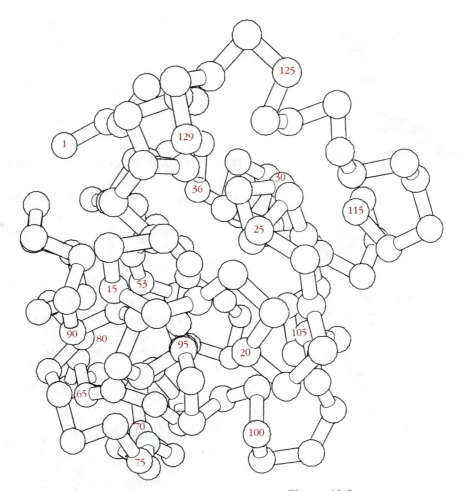

Amino acids	Rod length (cm)
6 to 127†	3.5
12 to 92	5.4
15 to 96*	8.4
20 to 99	6.9
21 to 104	5.9 (trim small clip on connector for 21)
26 to 120	3.0
31 to 110	4.3
46 to 109*	12.6
58 to 95	5.5
76 to 94†	3.6

Figure 10-3
α-Carbon drawing of lysozyme.

*Temporary support rod to be removed later. Mark with tape to remove later.
†Disulfide bonds.

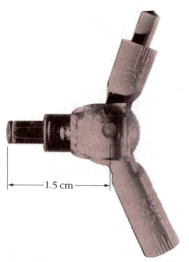

Figure 10-4
Amino acid with a connector cup, which will hold the side chain. The distance between the α- and β-carbons is 1.5 cm.

Once all the support rods are in place, they should be glued. Be careful to glue only the *permanent* rods, not the temporary rods that are marked with tape. After the glue has dried thoroughly, you may *carefully* remove the temporary support rods. Figure 10-2 shows the lysozyme model, and Figure 10-3 is an α-carbon representation.

Adding the Amino Acid Side Chains

LabQuip amino acid side chains will be attached to 19 amino acids in the active site. Blackwell model connectors will be used to connect the side chains to the completed model. The side chains listed in Table 10-2 will be added to the acitve site.

The precise position of the connectors on the α-carbons is set by triangulation. The end of the connector clip corresponds to the position of the β-carbon of the amino acid. The β-carbon is 1.5 cm from the center of the α-carbon sphere (Figure 10-4). All the triangulation distances are measured to the center of the α-carbon sphere. You will need to approximate the mea-

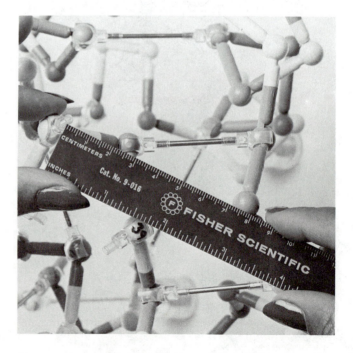

Figure 10-5
The position of a connector cup is set using triangulation distances. The distance is measured from the end of the cup to the center of a neighboring α-carbon.

surements; absolute precision is difficult and in fact un-necessary. Try to set the position of the β-carbon as close as possible to all the values you are given.

Place the connector clip on the appropriate α-carbon. Use one triangulation distance from Table 10-2 to get an idea of which way the connector will be pointing (Figure 10-5). This should indicate what position the connector should be in on the α-carbon (as you know, the con-nectors have two different-sized cup halves, and these must be positioned correctly to allow rotation on the α-carbon ball). Place a drop of glue on the α-carbon as you would to glue the connector. This allows the con-nector to be moved but prevents it from slipping. Now set the position of the connector using all the triangu-lation distances. You should recheck them all when you have finished setting all the distances. Apply more glue to permanently affix the connector.

Adding the Substrate Support Rods

The support rods for the substrate molecule are attached in a similar manner. Prepare the support rod using the rod length given in Table 10-3 and attach two connectors. Clip it onto the appropriate α-carbon, and set the posi-tion of the rod assembly as you did for the β-carbons of the amino acid side chains. Put a small drop of glue on the α-carbon to make the connector assembly movable but not loose. The triangulation distances are the dis-tances from the centers of the (free) connector cup spheres to the centers of the α-carbons listed. Again, these are approximate distances, but try to set them care-fully. Once the rod is positioned, hold it in place with one hand and apply some glue to the α-carbon. Hold it for a minute to make sure it will not slip. Be careful not to move the rod until it is thoroughly dry. Recheck the measurements to be certain they are correct. Figure 10-6 shows the model with support rods in place.

Before adding the amino acid side chains or the oli-gosaccharide substrate, you should study the model of lysozyme alone. Salient features of the protein structure are discussed later.

Table 10-2 Triangulation distances for positioning side chains to lysozyme

Side Chain	α-Carbon Number	Neighbor Distances (cm)							
		α-Carbon	Distance	α-Carbon	Distance	α-Carbon	Distance	α-Carbon	Distance
Glu	35	32	4.2	55	6.3	56	4.9	37	7.1
Ser	36	2	8.6	33	6.2	38	5.1	32	5.5
Asn	37	2	7.8	3	9.0	32	8.0	33	5.6
Asn	39	2	6.6	3	7.5	33	8.7	36	5.3
Asn	44	35	8.8	36	8.6	51	6.3	52	5.3
Asn	46	44	7.0	49	6.3	50	5.1	52	7.0
Asp	52	44	5.1	45	6.3	43	5.4	54	7.2
Leu	56	28	8.2	32	6.0	35	7.6	53	7.8
Gln	57	32	7.8	36	6.3	42	7.0	44	7.5
Arg	61	49	8.0	50	5.3	59	6.0	62	4.6
Trp	62	58	8.6	59	6.3	64	7.3	73	5.7
Trp	63	58	5.3	59	5.1	60	5.7	74	6.3
Leu	75	62	5.4	63	4.5	64	7.1	73	5.9
Asp	101	97	6.2	98	4.9	99	5.9	100	4.8
Asp	103	98	8.4	99	7.2	100	8.4	106	4.9
Ala	107	95	8.6	99	5.8	101	8.4	98	5.8
Trp	108	32	8.3	35	7.6	56	7.4	57	7.9
Val	109	35	8.4	106	7.8	111	7.0	112	6.7
Arg	114	30	8.0	34	6.2	35	8.7	110	5.1

Table 10-3 Triangulation distances for support rod connectors: lysozyme-NAM-NAG-NAM

Rod Length (cm)	Support Position (α-Carbon)	Neighbor Distances (cm)								Site	Sugar	Atom
		α-Carbon	Distance	α-Carbon	Distance	α-Carbon	Distance	α-Carbon	Distance			
6.2	Gly 102	107	6.2	62	7.6	101	7.9	103	7.7	B	NAM	O5
4.2	Ile 98	108	3.7	58	7.7	63	8.4	103	7.9	C	NAG	O3
4.0	Asn 59	58	7.1	52	7.6	57	7.8	—	—	D	NAM	O4

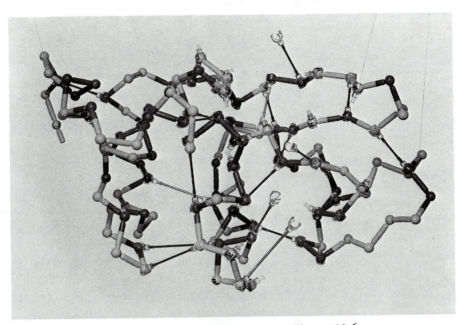

Figure 10-6
Support rods for the lysozyme substrate.

Saccharide Structure

Glucose and its derivatives are six-carbon carbohydrate molecules know as *aldoses*. These linear molecules cyclize in a hemiacetyl linkage to form six-membered rings called *pyranoses*. The pyranose ring is not planar but instead assumes a stable conformation known as the chair conformation (Figure 10-7). Four of the carbons in noncyclized glucose are chiral, and a fifth, C1, is made chiral by the cyclization reaction (it now has four bonds). After cyclization, the C-1 carbon is in either the α or the β conformation. In α-glucose the oxygen atom is below the plane of the ring, and in β-glucose the oxygen atom is within the plane of the ring. When this glucose is linked to another sugar at this carbon, the linkage is either α or β, depending on the position of the glycosidic oxygen.

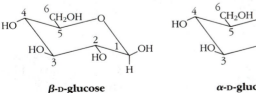
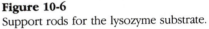

Figure 10-7
α- and β-D-glucose in the chair conformation. The carbons are numbered according to a standardized numbering system.

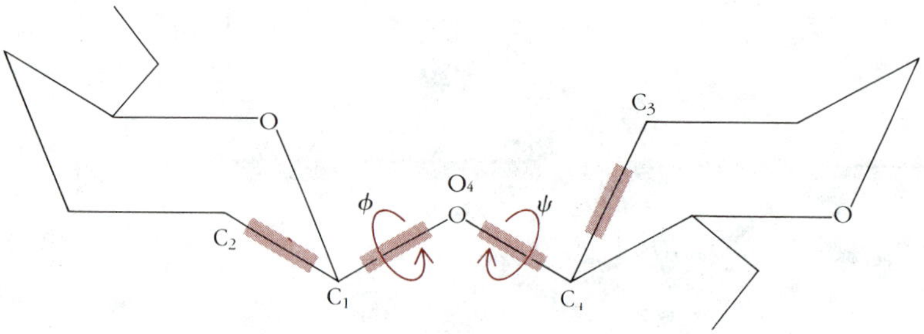

Figure 10-8
Four bonds are used in setting the torsional angles about the glycosidic link between sugars. These bonds are highlighted in the figure. The angles ϕ and ψ are indicated.

There is rotation about both bonds in the glycosidic linkage between two saccharides. The torsional angles that describe these rotations may be specified as for any other torsional rotations (Figure 10-8).

The hydroxyl groups on glucose may be derivatized by the addition of a variety of side chains. In the saccharides NAM and NAG, these side chains are a lactyl group on C-3 and an *N*-acetyl group on C-2, respectively. The position of the O or N linked to the ring carbon (C-3 or C-2, respectively) is fixed. However, there is rotation about the C-O or C-N bond, and a torsional angle may be used to set this rotation in the saccharide models. The numbering system for the substrate sugars is shown in Figure 10-9.

Assembly of the Lysozyme Substrate

You will construct a model of the pentasaccharide NAM-NAG-NAM-NAG-NAM. This molecule fills five (B, C, D, E, and F) of the six sugar binding sites A, B, C, D, E, and F) in the active site cleft of lysozyme. The atomic structure of the four sugars that bind in sites A,B, C, and D is known from x-ray crystallographic experiments, but the precise structure of the remaining sugars is not known and is predicted by model building.

The three-dimensional structure of NAM-NAG-NAM as bound to the B, C, and D sites of lysozyme was used to calculate the torsional angles about the bridge oxygen atoms that link C-1 of one sugar to C-4 of the next. These angles, designated ϕ and ψ, are defined in Figure 10-8. Table 10-4 has the values for these two pairs of angles.

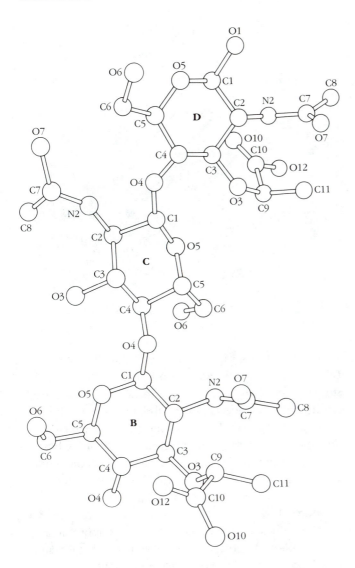

Figure 10-9
NAM-NAG-NAM trisaccharide in the
conformation found in the lysozyme-
substrate complex. The atomic numbering
scheme is shown.

Table 10-4 ϕ, ψ Angles between saccharides for
NAM-NAG-NAM in complex with lysozyme

Defining Atoms	Site Location	Angle (°)
C2-C1-O4-C4	B-C	147
C1-O4-C4-C3	B-C	137
C2-C1-O4-C4	C-D	139
C1-O4-C4-C3	C-D	97

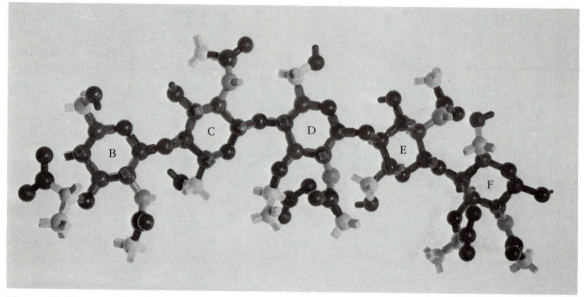

Figure 10-10

NAM-NAG-NAM-NAG-NAM pentasaccharide that lies in the B-C-D-E-F sugar binding sites of lysozyme.

Assemble the LabQuip model parts to build the five β-linked glucose units as shown in Figure 10-10. Put the sugars in the chair conformation as shown in Figure 10-7. Do not add the side chains to the sugars until the torsional angles are set. The torsional angles for the D-E and E-F links are unknown and can be adjusted for the best fit after the ligand model is attached to the lysozyme model. The other torsional angles that determine the conformation of the trisaccharide are known.

Figure 10-9 shows the conformation of this trisaccharide. The angles ϕ and ψ can be set approximately to the values in Table 10-4 by referring to Figure 10-11 which precisely defines these torsional angles. To set the angles on the LabQuip models, the four atoms should be located and viewed down (from 2 to 3) the bond from atom 2 to 3. The model is positioned so that the bond between atoms 3 and 4 is horizontal (Figure 10-11*c*). Then rotate bond 1 to 2 counterclockwise (by applying torsion to the 2 to 3 bond) until the ϕ (or ψ) angle desired is set. A protractor (or pair of completed backbone units set into a ∪ shape) can be used to estimate the numerical value. The value 0° is defined in Figure 10-11*d* when atom 1 eclipses atom 4 as shown. If you use model units as a protractor to measure torsional angles, make

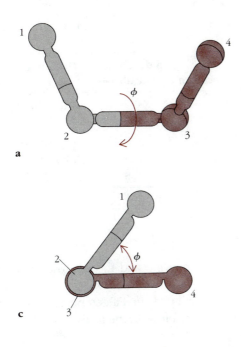

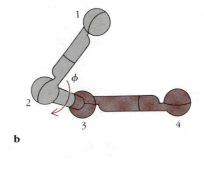

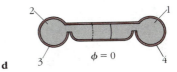

Figure 10-11
Setting torsional angles. **a**, **b**, and **c** show the bonds used to set the angles θ. **d** depicts an angle $\theta = 0$. Note that the bonds are eclipsed.

sure that when the units are in the conformation shown in Figure 10-11*d*, the torsional angle between them reads 0, not 180.

Construct the 3-lactyl and 5-*N*-acetyl side chains (Figure 10-9), and make sure the chiral carbon, C-9, in the lactyl side chain is the correct handedness. Attach these side chains to the oligosaccharide, being careful not to disturb the torsional angles you have just set. Each NAM residue (B, D, and F) has a lactyl group attached to the C-3 atom. All five residues have an *N*-acetyl group attached in the C-2 position. Set the torsional angles for the side chains on sugars NAM-NAG-NAM (B, C, and D) using the values in Table 10-5. The conformation of the B, C, and D sugars should match Figure 10-9.

Discussion of Lysozyme Structure

Lysozyme is a small molecule with the dimensions of 45 Å × 30 Å × 27 Å. Its shape is ellipsoidal, with a deep gash running along one face. This noticeable feature is the enzyme's *active site* where the oligosaccharide substrate

Table 10-5 Torsional angles for side chains in the trisaccharide NAM-NAG-NAM

Site	Location	Side Chain Location	Defining Atoms	Angle (°)
NAM	B	C-2	C1-C2-N2-C8	169
		C-3	C2-C3-O3-C9	14
NAG	C	C-2	C1-C2-N2-C8	93
NAM	D	C-2	C1-C2-N2-C8	93
		C-3	C2-C3-O3-C9	285

binds to the enzyme and the catalytically active amino acid side chains are found. Lysozyme's active site is large for such a small protein. Fifteen percent (a sizeable fraction) of the protein is in the active site. The active site is cavernous and appears to be deep enough to surround the substrate.

You will note that the surface of the lysozyme model is composed of primarily red backbone units and that green is found in the interior of the model. Lysozyme is a good example of the feature common to globular proteins: a hydrophilic surface surrounding a hydrophobic core.

Many examples of α helices, β strands, and turns are obvious in the lysozyme model. Alpha helices are found in positions R5 to H15, L25 to E35, C80 to L84, and T89 to K96. These helical segments were probably obvious to you as you were constructing the model. There are two regions of β sheet in lysozyme. One is very short and is composed of K1 to F3 paired antiparallel with F38 to T40. The second β sheet lines one face of the active site. Amino acids A42 to N46, S50 to G54, and Q57 to S60 make up this antiparallel sheet.

Turns are found joining all the aforementioned segments of secondary structure and in a few other places. Type I turns are found at G4 to E7 and G54 to Q57. The latter turn is a rare example of a buried turn (see p.86 for a discussion of buried turns). This one is found deep in the active site. Type II turns are found at L17 to Y20, N19 to G22 (these turns overlap and double back on each other), D103 to N106, C115 to T118, and I124 to C127. Type III turns occur at positions K13 to G16, N59 to W62, T69 to S72, and N74 to N77.

Adding Side Chains to the Lysozyme Model

You should now assemble the 19 amino acid side chains using the LabQuip model parts and amino acid structures in Table 10-6. These side chains should be snapped onto the connector cups on the α-carbons in the appropriate positions. Once the substrate model has been constructed and attached to the enzyme model, you should attempt to position the side chains so that they make appropriate contacts with the substrate. Refer to Figure 10-12 to help position the side chain torsional angles. Figure 10-12 is an α-carbon representation of lysozyme showing the active-site side chains.

Adding the Substrate to the Lysozyme Model

The completed oligosaccharide should be set into the active site of the lysozyme model. Hold the oligosaccharide so that (B) NAM O5, (C) NAG O3, and (D) NAM O4 are positioned over the three support rods. Gently snap the oxygen atoms into the connector cups without changing the torsional angle settings between the sugars. The positions of the E and F sugars, as well as the positions of their side chains, may be adjusted so that the oligosaccharide fits neatly into the active site. You may want to construct a support rod to hold the F sugar (NAM) in an approximate position to aid your observation of the model. A 4 cm support rod may be cut and used to connect amino acid T 43 to O4 of (F) NAM. Since this is an estimated distance, you do not need to glue the support rod.

Table 10-6   Materials list for lysozyme: NAM-NAG-NAM-NAG-NAM complex

Protein Side Chains	Name of Atom Type	Color	Part Number	Number
ALA (107)	Tetrahedral carbon	White	3241	1
VAL (109)	Tetrahedral carbon	White	3241	3
LEU (56, 75)	Tetrahedral carbon	White	3241	8
TRP (62, 63, 108)	Tetrahedral carbon	White	3241	3
	Indole ring	Purple	3385	3
SER (36)	Tetrahedral carbon	White	3241	1
	Hydroxyl oxygen	Red	3574	1
ASN (37, 39, 44, 46)	Tetrahedral carbon	White	3241	4
	Peptide	Blue	3313	4
		Blue	3323	4
GLN (57)	Tetrahedral carbon	White	3241	2
	Peptide	Blue	3313	1
		Blue	3323	1
ASP (52, 101, 103)	Tetrahedral carbon	White	3241	3
	Carboxylate	Red	3494	3
GLU (35)	Tetrahedral carbon	White	3241	2
	Carboxylate	Red	3494	1
ARG (61, 114)	Tetrahedral carbon	White	3241	6
	Guanidinium	Blue	3353	2

Total (Lysozyme Side Chains)

Name	Color	Part Number	Quantity
Tetrahedral carbon	White	3241	33
Indole Rings	Purple	3385	3
Hydroxyl	Red	3574	1
Carboxylate	Red	3494	4
Peptide	Blue	3313	5
Guanidinium	Blue	3353	2

Backbone Units

Color	Quantity
Red	63
Green	66
Connectors	79

Table 10-6   (continued)

Saccharides				
Saccharide/Number	Name of Atom Type	Color	Part Number	Number
NAM 3	Tetrahedral carbon	White	3241	24
	Tetrahedral carbon (double, short-arm)	White	3261	3
	Hydroxyl oxygen	Red	3574	3
	Ether oxygen	Red	4544	7
	Carboxylate	Red	3494	3
	Ether oxygen	Red	4564	3
	Peptide bond	White	3281	3
NAG 2	Tetrahedral carbon	White	3241	12
	Tetrahedral carbon (double socket, short-arm)	White	3261	2
	Ether oxygen	Red	4544	6
	Hydroxyl oxygen	Red	3574	4
	Peptide bond	White	3281	2
	Ether oxygen	Red	4564	2

Structural Features of the Lysozyme-Substrate Complex

The proposed structure of the enzyme substrate complex was originally deduced by a series of experiments. The use of many substrate analogs was a critical feature of this research. Initial x-ray crystallographic analysis of the complex of lysozyme and an inhibitor, NAG-NAG-NAG or tri-NAG (or NAG_3), yielded detailed information about the dimensions of the active site and the intimate interaction of this inhibitor with lysozyme. Tri-NAG appeared to fill about half of the active site. It appeared that the real substrate for the enzyme was a six-sugar segment of the bacterial cell wall. However, complexes of the enzyme and hexa-NAG (NAG_6) were not stable, since the sugar acted as a substrate and was hydrolyzed by the enzyme. The positions of the other three sugars were first predicted by model building. The positions of the six sugars were named A, B, C, D, E, and F as indicated in Figure 10-13.

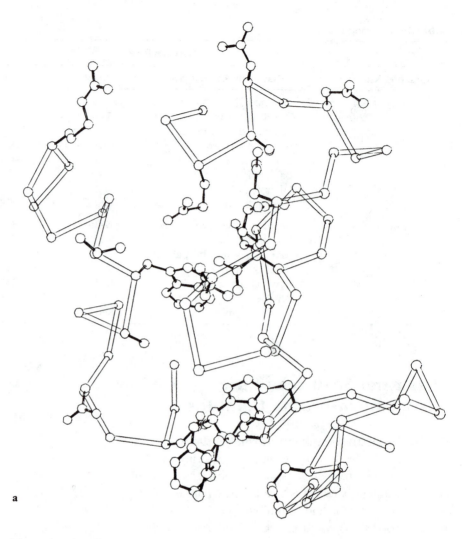

a

Figure 10-12

α-Carbon representation of the active site of lysozyme showing the positions of the side chains.

Which bond of the six-sugar segment cleaved by lysozyme remained to be determined. Since tri-NAG was stable to lysozyme, the cleaved bond was neither the A-B nor B-C bond. It was known that bacterial cell wall oligosaccharides are an alternating polymer of NAG and NAM. The sugar NAM does not fit into the C site of the enzyme because of the bulky lactyl group (this will be observed on the completed model). The hexasaccharide must have the sequence NAG-NAM-NAG-NAM-NAG-NAM in the six positions A-B-C-D-E-F. It was also known

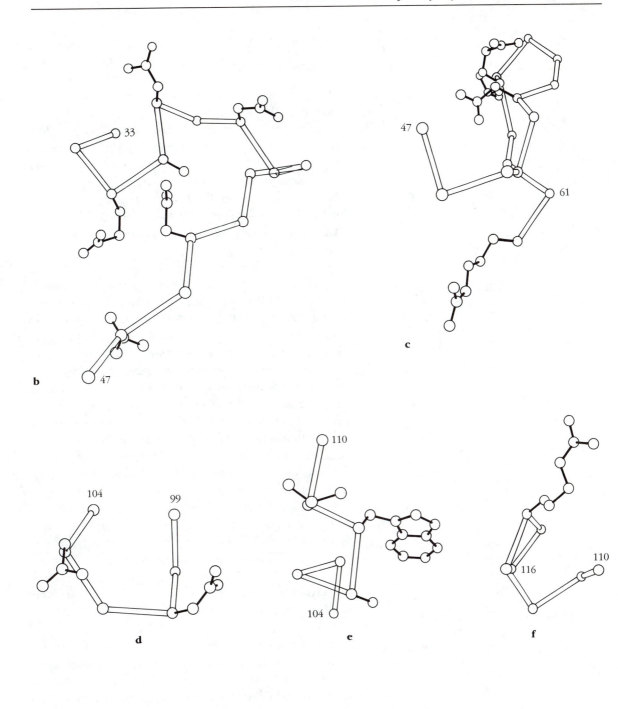

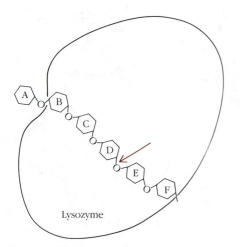

Figure 10-13
Cartoon of a six-unit sugar bound in the active site of lysozyme. The positions of the sugar subsites are indicated by the letters A, B, C, D, E, and F. The arrow indicates the site where the sugar is cleaved by the enzyme.

that the NAM-NAG bond was cleaved. This places the site of cleavage unequivocally at the D-E glycosidic bond.

There are six subsites on lysozyme for binding a hexasaccharide. If you examine the model, you will see that few side chains are at the A subsite, and A shows a low affinity for sugar. The x-ray analysis of a tetrasaccharide (NAG$_4$) that binds to the sites ABCD shows the interaction of the sugar with the enzyme to be primarily van der Waals interactions. No details of binding are established for the E or F subsites, and the positions of the saccharides in these positions are deduced from model building.

It has been proposed that the D subsite has low affinity for saccharides in the chair conformation, based on binding studies. The model shows this site to be fully occupied in the NAM-NAG-NAM complex with lysozyme. The explanations for this paradox follow. First, the D sugar is likely not in exactly the position that corresponds to the position for which the binding affinity measurements were made. Second, the estimation of binding energies was made with a mixture of α and β anomers, α being present at 66%. Lysozyme binds the β anomer exclusively. Studies using a tetrasaccharide inhibitor whose D sugar was in a half chair conformation showed that, in fact, this inhibitor makes significantly better atomic contacts with the enzyme, penetrating on average 0.5 Å deeper into the active site. The D sugar of NAG-NAG-NAG-N-acetyl-lactone inhibitor is in a similar conformation to that proposed by investigators as being the actual conformation taken by the substrate in the transition state that occurs during catalysis. The NAG$_3$ N-acetyl-lactone inhibitor is called a *transition state analog* and is an example of a powerful class of enzyme inhibitors (Figure 10-14).

A conclusion from the binding-affinity data and the structural analysis of the complex is that the sugar ring in the D subsite is not distorted from the chair conformation when it initially binds to the enzyme. This distortion must occur later during hydrolysis when the reaction intermediate is formed. (For further discussion, see the Suggested Readings at the end of this chapter.)

Active-Site Catalytic Side Chains

Knowledge of the details of interaction of lysozyme and its substrate and finding the site of enzyme action on the sugar allowed investigators to identify the amino acid residues that are involved in catalysis. These residues are called *catalytic groups* and are residues glutamic acid 35 and aspartic acid 52. Find these residues on your model. These two side chains are positioned on opposite sides of the active site and straddle the NAM-NAG β, 1-4 glycosidic bond. This is important for the actual reaction mechanism. You should notice that Glu 35 is in a nonpolar environment, surounded by hydrophobic side chains (Ile 55, Val 56, Trp 108, Val 109). Asp 52 is in a polar environment and is surrounded by charged side chains. Lysozyme, like most enzymes, has an optimum pH for activity. At pH 5.5, it is likely that Glu 35, with its nonpolar surroundings, is protonated and uncharged and that Asp 52 is ionized and negatively charged. A simplified version of the actual reaction mechanism is that glutamic acid 35 donates its proton to break the β 1-4 glycosidic bond. This leaves the disaccharide of E and F and a carbonium ion C-1 on the D sugar. The presence of the positively charged carbonium ion is stabilized by the negative charge on aspartic acid 52. The carbonium ion then reacts with OH from the solvent, and the reaction is complete.

NAM-NAG-NAM Conformation

As stated previously, the conformation of oligosaccharides can be described by specification of the ϕ and ψ torsional angles about the bridging oxygen atom in the β 1-4 linkage. The sugar groups, although slightly flexible, are fixed with the pyranose ring in the chair conformation.

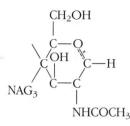

Carbonium derivative of tetra-NAG

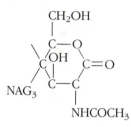

Lactone analog of tetra-NAG

Figure 10-14
The structure of the transition state analog NAG$_3$ *N*-acetyl lactone is similar to the carbonium ion reaction intermediate.

Two important determinants of the conformation of oligosaccharides in solution are steric accommodation of successive sugar units and hydrogen bonds that may form between hydroxyl groups of the sugar groups adjacent in sequence. The complex here shows the B, C, D trisaccharide embedded in the active site cleft. An important question is whether the conformation of the trisaccharide changes after binding to the lysozyme, and the detailed answer to this is not known. For saccharides in general, the ϕ, ψ angles in the enzyme-ligand complex are within 10° to 20° of expected values based on measurements of these angles in solution or single crystals of the saccharides without the enzymes. In general, ligands do not suffer radical changes in conformation when they bind to proteins.

Interactions Between NAM-NAG-NAM and Lysozyme

The enzyme changes its conformation on binding the substrate analog. Atomic coordinates for lysozyme alone, *not* the enzyme-substrate analog complex, were used in building this model (the atomic coordinates for the complex are not known). The major movements are Trp 62 toward site B, and Trp 63 toward site C. The main chain and side chains for residues Gly 71 through Leu 75 are also shifted. The fit that occurs between the enzyme and substrate is closer than depicted here. This fit is described in Table 10-7, which lists hydrogen bonds between enzyme and substrate. In addition, many van der Waals contacts are made in the enzyme-substrate complex. These are shown in Table 10-8. Compare the model and adjust the side chain conformations to observe these interactions.

As expected, the enzyme-ligand complex is a close-packed structure. The interface excludes water (the x-ray analysis shows that ten water molecules in the active site region are displaced when the ligand binds). Figures 10-15 to 10-17 show space-filling models of the ligand, the enzyme, and the complex.

Table 10-7   Possible hydrogen bonds between NAM-NAG-NAM and lysozyme*

Protein	Amino Acid	Atom	Site	Substrate Atom	Distance (Å)
Asp	103	OD	B	O(10)	2.9
Asp	101	OD	B	O(6)(+H_2O)	3.7
Trp	63	NE	C	O(3)(+H_2O)	4.0
Trp	62	NE	C	O(6)	3.0
Asn	59	N	C	O(7)(+H_2O)	4.2
Asp	52	OD	D	O(5)	3.5
Val	109	N	D	O(6)	3.4
Asn	46	ND	D	O(10)	3.0
Glu	35	OE	D	O(6)	2.8

*Distances to 4.2 Å were included. It is not certain whether the longer separations correspond to hydrogen bonds, although they may form by side chain rotation or via the presence of an intervening water molecule that bridges the distance. (See p. 185 for a description of the atomic numbering notation.

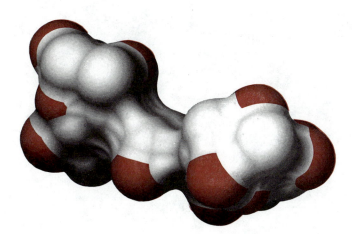

Figure 10-15
Space-filling representation of the lysozyme substrate. Refer to Figure 2-9 for the coloring scheme and details of the representation.

Figure 10-16
Space-filling representation of lysozyme.
Note the deep groove on the left side.

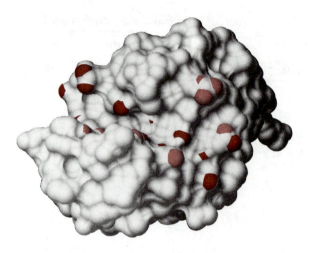

Figure 10-17
Space-filling representation of the lysozyme
NAM-NAG-NAM complex. Note that the
structure is close-packed. The substrate is
shaded and sits in the active site of the
unshaded enzyme. Note there is still a deep
unfilled pocket in the enzyme's active site.

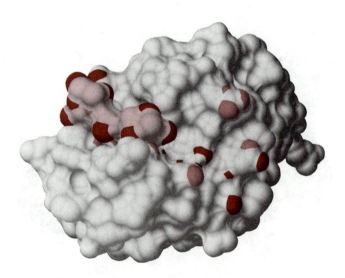

The amino acid numbering scheme used in Tables 10-7 and 10-8 is based on the Greek alphabet. N, CA, C, and O correspond to backbone atoms N, Cα, C′, and the carbonyl oxygen. The side chains are labelled going outward from Cα. The positions, in order, are α, β, γ, δ, ε, and ζ, which translate to A, B, G, D, E, and Z. In our notation, a given atom is indicated by the atom type (for example, C, N, O) followed by its position (A, B, G, and so on). In cases where the side chain branches, the two atoms are numbered 1 or 2. For most amino acids the numbering scheme is straightforward. Tryptophan is shown in Figure 10-18.

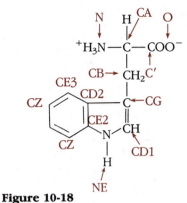

Figure 10-18
Tryptophan. The atomic numbering scheme used in Tables 10-7 and 10-8 is shown.

Table 10-8 Van der Waals contacts between lysozyme and NAM-NAG-NAM*

NAM 1 C6	Trp 63			
	CZ 3.9			
NAM 1 O6	Trp 63	Asp 103		
	CZ 3.9	OD 4.1		
NAM 1 C10	Asp 103	Asp 103	Asp 103	
	CG 3.8	OD 3.5	OD 3.3	
NAM 1 O12	Asp 103			
	OD 3.9			
NAG 2 C1	Ala 107			
	O 3.3			
NAG 2 C2	Ala 107			
	C 4.0			
NAG 2 C3	Ala 107	Ala 107	Ala 107	Ala 107
	CA 3.9	C 3.9	O 3.1	CB 3.9
NAG 2 O3	Trp 63	Ala 107	Ala 107	Ala 107
	NE 4.0	CA 4.0	O 3.7	CB 3.5
NAG 2 O6	Trp 62	Trp 62		
	CD1 3.9	CE 4.0		
NAG 2 N2	Ala 107	Ala 107	Ala 107	Trp 108
	CA 3.9	C 3.1	CB 3.9	N 4.1
NAG 2 C7	Ala 107	Trp 108	Trp 108	
	C 3.8	CD1 4.1	NE 4.1	

*The sugar positions 1, 2, and 3 correspond to B, C, and D.

Table 10-8   (continued)

NAG 2 C8	Leu 56 O 4.1	Gln 57 O 3.7	Ala 107 C 4.1	Ala 107 O 3.3
	Trp 108 CA 4.1	Trp 108 CG 3.4	Trp 108 CD2 3.8	Trp 108 CE 3.5
NAG 2 O7	Trp 63 CD1 3.9	Trp 63 NE1 3.6	Ala 107 O 4.0	
NAM 3 C1	Asn 46 ND 4.1	Asp 52 OD 3.6	Val 109 CG 3.4	
NAM 3 C2	Val 109 CG 3.5			
NAM 3 C4	Val 109 CG 3.9			
NAM 3 C5	Asp 52 OD 3.4	Val 109 CG 3.7		
NAM 3 C6	Glu 35 OE 4.0	Gln 57 O 4.0	Ala 107 O 3.8	Trp 108 CA 4.1
	Trp 108 CD1 4.2	Val 109 N 4.2	Val 109 CG 3.8	
NAM 3 O1	Val 109 CG 3.2			
NAM 3 O4	Ala 107 O 3.9			
NAM 3 O6	Glu 35 CD 3.4	Glu 35 OE 3.7	Trp 108 CA 3.6	Trp 108 C 3.9
	Trp 108 CD1 4.0	Val 109 N 3.4	Val 109 CG 3.1	
NAM 3 N2	Asn 46 ND 4.1			
NAM 3 C10	Asn 46 CG 4.0	Asn 46 ND 3.9		
NAM 3 O10	Asn 46 CB 4.0	Asn 46 CG 3.4	Asn 46 OD 4.0	Asn 46 ND 3.0
NAM 3 O12	Asn 46 CB 3.8	Asn 46 CG 4.0	Asn 46 ND 4.0	

Summary

The antibacterial property of the enzyme lysozyme resides in the ability of the enzyme to catalyze the hydrolysis of the bonds in bacterial cell walls. Lysozyme has a binding site that recognizes and is specific for five of the sugar chains found in the cell wall. The sugars fit into the binding site in only one way. The binding interactions, which are van der Waals contacts and hydrogen bonds, yield a close-packed structure for the complex.

The enzyme is approximately ellipsoidal in shape, with its active site apparent as a deep cleft when sugars are not bound. About 15% of the side chains are directly involved in the binding of substrate and hydrolysis of the sugar-sugar bond.

Four disulfide bonds stabilize this protein of 129 amino acids. The molecule is constructed from α helices and β strands that are clustered in separate regions of the molecule.

Suggested Readings

Lysozyme reviews

Phillips DC: The three-dimensional structure of an enzyme molecule. *Sci Am* 1966; 215:75–90. Review of lysozyme structure.

Imoto T, Johnson LN, North ACT, et al: Vertebrate lysozymes, in Boyer PD (ed): *The Enzymes,* ed 3. New York, Academic Press Inc, vol 7, 1971, pp. 666–868. Thorough discussion of lysozyme structure and function.

Lysozyme function

Dahlquist FW, Rand-Meir R, Raftery MA: Demonstration of carbonium ion intermediate during lysozyme catalysis. *Proc Natl Acad Sci USA* 1968; 61:1194–1198. Discussion of lysozyme reaction mechanism.

Kelly JA, Sielecki AR, Sykes BD, James MNG, Phillips DC: X-ray crystallography of the binding of the bacterial cell wall trisaccharide NAM-NAG-NAM to lysozyme. *Nature* 1979; 282:875–878. X-ray diffraction analysis of a lysozyme sugar complex. The details of the interactions that stabilize the complex are presented.

Oligosaccharide conformation

Goldsmith E, Fletterick RJ: Oligosaccharide conformation and protein saccharide interactions in solution. International Carbohydrate Symposium, Vancouver, BC, in Marchessault RH (ed): *Pure and Applied Chemistry*, New York, Pergamon Press, vol 55, 1983, pp. 577–588. Discussion of carbohydrate conformation in solution and on the enzyme glycogen phosphorylase.

MOLECULAR MODELS AND PARTS

With the 136 model parts provided in the *Blackwell Molecular Models* kit, you can build several different models, both temporary and permanently glued. For example, you may build permanently glued models of the following:

- *cro* repressor

- Lysozyme

- tRNA

- tRNA and β sheet

- A-, B-, or Z-DNA and β sheet

- A-, B-, or Z-DNA and insulin

- A-, B-, or Z-DNA and collagen

- Insulin, collagen, and β sheet

The following table lists the parts and time needed to construct the various structures described in this text.

| Model Name | Glued | Parts Required | | | | Construction Time |
		Red	Green	Connectors	Rods (cm)	
α helix	No	19	0	0	0	10 minutes
3₁₀ helix	No	19	0	0	0	10 minutes
Collagen	Yes	12	24	22	43	45 minutes
β sheet	Yes	24	0	18	25	30 minutes
Type 1 turn	No	4	10	0	0	10 minutes
Type 2 turn	No	4	10	0	0	10 minutes
Type 3 turn	No	4	10	0	0	10 minutes
B-DNA	Yes	24	24	48	210	1 hour
A-DNA	Yes	24	24	48	250	1 hour
Z-DNA	Yes	24	24	48	135	2 hours
tRNA	Yes	6	70	54	211	2 hours
Insulin	Yes	25	26	12	22	1 hour
cro Repressor	Yes	66	66	24	50	$2\frac{1}{2}$ hours
Lysozyme With side chains and substrate*	Yes	63	66	79	105	$2\frac{1}{2}$ hours 5 hours
Total		318	354	353	1051	19 hours
Less temporary models		268	324			

*Available from Lab Quip.

*Italicized page numbers indicate figure.